COPPER ALLOYS – EARLY APPLICATIONS AND CURRENT PERFORMANCE – ENHANCING PROCESSES

Edited by **Luca Collini**

Copper Alloys – Early Applications and Current Performance – Enhancing Processes
Edited by Luca Collini

Published by InTech
Janeza Trdine 9, 51000 Rijeka, Croatia

Publishing Process Manager Iva Simcic

Technical Editor Teodora Smiljanic

Cover Designer InTech Design Team

Additional hard copies can be obtained from orders@intechopen.com

Copper Alloys – Early Applications and Current Performance – Enhancing Processes,
Edited by Luca Collini
p. cm.
ISBN 978-953-51-0160-4

For EU product safety concerns:
IN TECH d.o.o., Prolaz Marije Krucifikse Kozulić 3, 51000 Rijeka, Croatia,
info@intechopen.com or visit our website at intechopen.com.

Meet the editor

Dr. Eng. Luca Collini is researcher in "Mechanical design and construction of machines" since 2005, at the Department of Industrial Engineering, University of Parma, Italy. As a lecturer of the Faculty of Engineering of Parma University, Dr. Eng. Collini teaches courses in the construction of machinery, technical drawing, and product design and development for mechanical and management engineering students. At the present he is involved in the Socrates/Erasmus European Community program for the international mobility of students. He is author of over 70 journal papers and contributions of national and international conference proceedings, his research activity embraces the mechanical behavior of materials, in particular fatigue and fatigue crack resistance of metals, the advanced mechanical design methods, and the structural integrity assessment techniques based on vibration analysis.

Contents

Preface

The use of copper and its alloys by humans goes back thousands of years. Probably copper is the most ancient metal extracted and processed by mankind. With good strength and ductility, ease of processing and good availability, copper has been for millennia an excellent material for creating objects, jewelry and tools of daily use or components with high technological content.

Nowadays, copper is used in its pure state mostly in applications that require high heat or electric conductivity, while its alloys (bronzes and brasses) are widely used in several fields thanks to their excellent corrosion and wear resistance. Copper and its alloys have been and still are a great natural resource for the growth of mankind.

This book collects, without the claim to be exhaustive, the most recent contributions of engineers and scientists who perform research on this material. These contributions cover two main topics: the development and application of advanced techniques of analysis of ancient relics made of copper, and the application of recently studied alloys, enhanced in mechanical and environmental performances. In my opinion the contrast of this two topics is fascinating in its own way.

In order to facilitate its reading, the book has been rationally divided into three distinct sections. The first section contains two papers, the first on the classification of copper and its alloys according to the modern systems, the second on the basic mechanism of interaction of copper with hydrogen.

The second section includes four contributions, two on the properties of alloys specifically developed for high-performance electromechanical connectors and for sliding elements, and two on the study of static and fatigue properties of copper subjected to a severe plastic deformation process.

The third section includes two very interesting contributions on the archaeometallurgy of copper.

Sure that all these high-level contributions are of interest, I would like now to extend my sincere thanks to all those who contributed to the drafting of this book, and in particular to prof. R. Konecna and dr. L. Kunz for their kind assistance.

Dr. Luca Collini,
Department of Industrial Engineering, University of Parma
Italy

Part 1

Introduction to the Copper Alloys

1

Copper and Copper Alloys: Casting, Classification and Characteristic Microstructures

Radomila Konečná and Stanislava Fintová
University of Žilina
Slovak Republic

1. Introduction

1.1 Copper

Copper is non-polymorphous metal with face centered cubic lattice (FCC, Fig. 1). Pure copper is a reddish color (Fig. 2); zinc addition produces a yellow color, and nickel addition produces a silver color. Melting temperature is 1083 °C and density is 8900 $kg.m^{-3}$, which is three times heavier than aluminum. The heat and electric conductivity of copper is lower compared to the silver, but it is 1.5 times larger compared to the aluminum. Pure copper

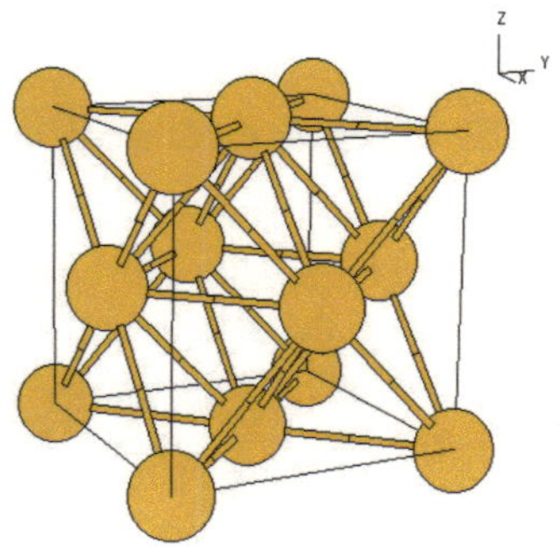

Fig. 1. FCC lattice (http://cst-www.nrl.navy.mil/lattice/struk/a1.html)

Fig. 2. Natural copper (http://jeanes.webnode.sk/prvky/med/)

electric conductivity is used like a basic value for other metals evaluation and electric conductivity alloys characterization (Skočovský et al., 2000, 2006). Copper conductivity standard (IASC) is determined as 58 Mss^{-1}. The pure metal alloying decreased its conductivity (Skočovský et al., 2000).

Before the copper products usage it has to pass through a number of stages. When recycled, it can pass through some stages over and over again. In nature, copper in its pure metal form occurs very often. Metallurgically from ores, where the chemical compound of copper with oxygen, sulphur or other elements occurs, copper is produced:

- chalcopyrite ($CuFeS_2$) – contains around 34.5 % of copper;
- azurite ($Cu_3[OH-Co_3]_2$) and malachite ($Cu_2[OH-Co_3]_2$) – alkaline copper carbonates;
- cuprite (Cu_2O) – copper oxide.

The beginning for all copper is to mine sulfide and oxide ores through digging or blasting and then crushing these to walnut-sized pieces. Crushed ore is ball or rod-milled in large, rotating, cylindrical machines until it becomes a powder, usually containing less than 1 % copper. Sulfide ores are moved to a concentrating stage, while oxide ores are routed to leaching tanks.

Minerals are concentrated into slurry that is about 15 % of copper. Waste slag is removed. Water is recycled. Tailings (left-over earth) containing copper oxide are routed to leaching tanks or are returned to the surrounding terrain. Once copper has been concentrated, it can be turned into pure copper cathode in two different ways: leaching & electrowinning or smelting and electrolytic refining.

Oxide ore and tailings are leached by a weak acid solution, producing a weak copper sulfate solution. The copper-laden solution is treated and transferred to an electrolytic process tank. When electrically charged, pure copper ions migrate directly from the solution to starter cathodes made from pure copper foil. Precious metals can be extracted from the solution.

Several stages of melting and purifying the copper content result, successively, in matte, blister and, finally, 99% pure copper. Recycled copper begins its journey to finding another use by being remelted. Anodes cast from the nearly pure copper are immersed in an acid bath. Pure copper ions migrate electrolytically from the anodes to "starter sheets" made from pure copper foil where they deposit and build up into a 300-pound cathode. Gold, silver and platinum may be recovered from the used bath.

Cathodes of 99.9% purity may be shipped as melting stock to mills or foundries. Cathodes may also be cast into wire rod, billets, cakes or ingots, generally, as pure copper or alloyed with other metals (http://www.mtfdca.szm.com/subory/med-zliatiny.pdf, http://www.copper.org/education/production.html).

Coppers mechanical properties (Fig. 3) depend on its state and are defined by its lattice structure. Copper has good formability and toughness at room temperature and also at reduced temperature. Increasing the temperature steadily decreases coppers strength properties. Also at around 500 °C the coppers technical plastic properties decrease. Due to this behavior, cold forming or hot forming at 800 to 900 °C of copper is proper. Cold forming increases the strength properties but results in ductility decreasing. In the as cast state, the copper has strength of 160 MPa. Hot rolling increases coppers strength to 220 MPa. Copper has a good ductility and by cold deformation it is possible to reach the strength values close to the strength values of soft steel (Skočovský et al., 2000, 2006).

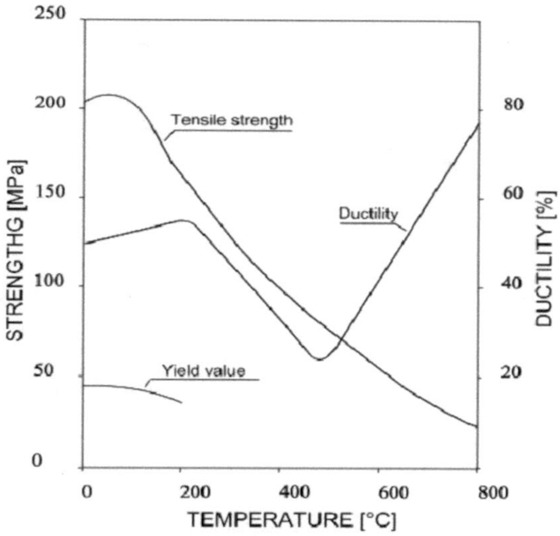

a) temperature influence on tensile strength, yield value and ductility

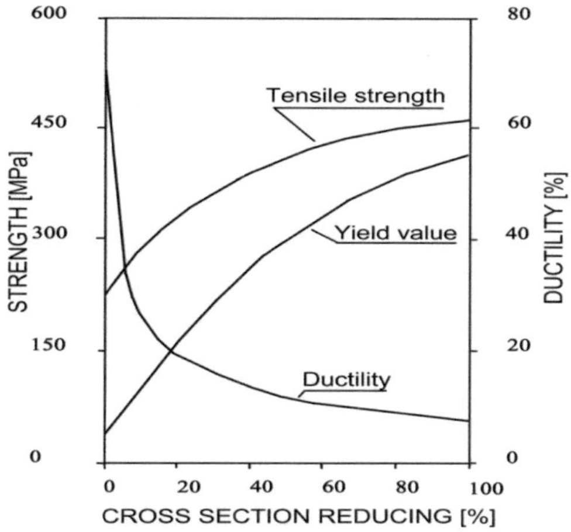

b) change properties due to the cold forming

Fig. 3. Copper mechanical properties (Skočovský et al., 2006).

Copper resists oxidation, however, it is reactive with sulphur and its chemical compounds, and during this reaction copper sulphide is created. Besides oxygen, the main contaminant, phosphor and iron are the significant copper contaminants. It is difficult to cast pure copper because large shrinkages during the solidification occur (1.5 %), and is the dissolving of a large amount of gasses at high temperatures disengaged during the solidification process and resulting in the melted metal gassing and the casting porosity (Fig. 4a, b). Cast copper microstructure is formed by non-uniform grains with very different sizes. Wrought copper microstructure consists of uniform polyhedral grains with similar grain size and it is also possible to observe annealing twins (Fig. 4c-e). Because of coppers reactivity, the dangers of surface cracking, porosity, and the formation of internal cavities are high.

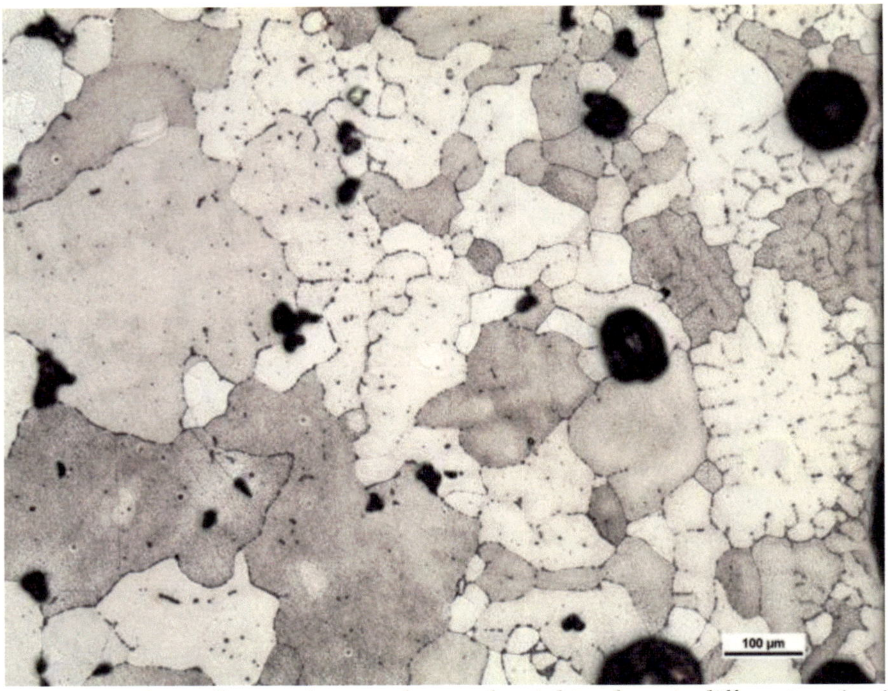

a) microstructure of cast Cu; specimens edge on the right, where is different grain size due to high cooling rate on the surface

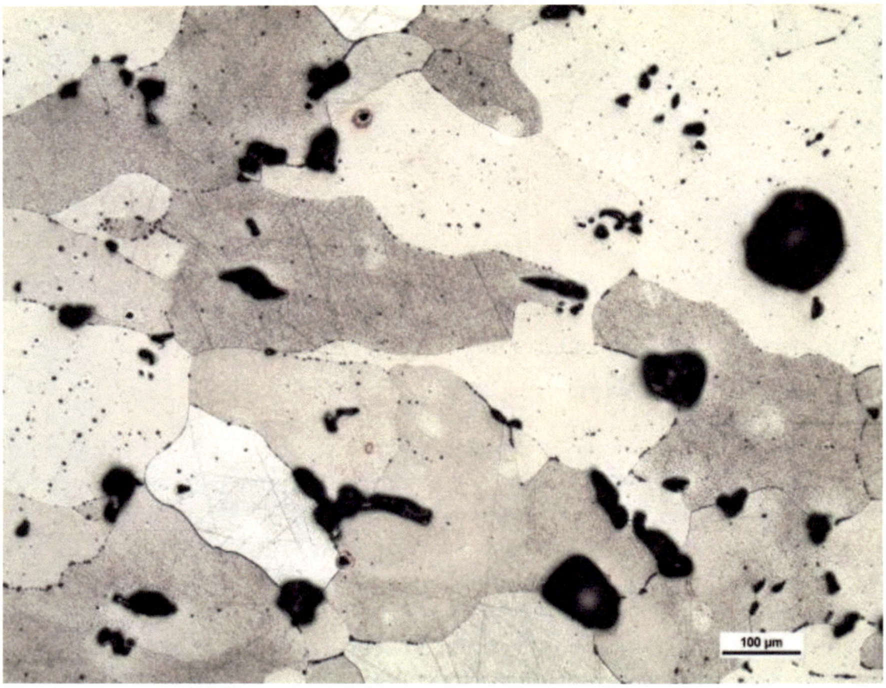

b) cast Cu; specimens middle part with big grains at low cooling rate

Fig. 4. Pure copper, chemically polished

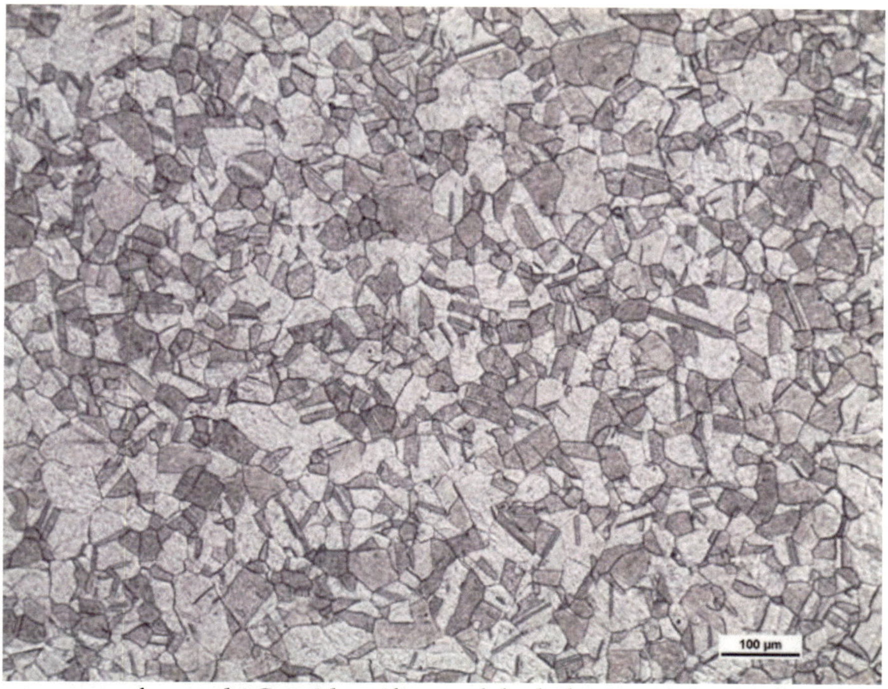

c) microstructure of wrought Cu with uniform polyhedral grains and annealing twins, white light

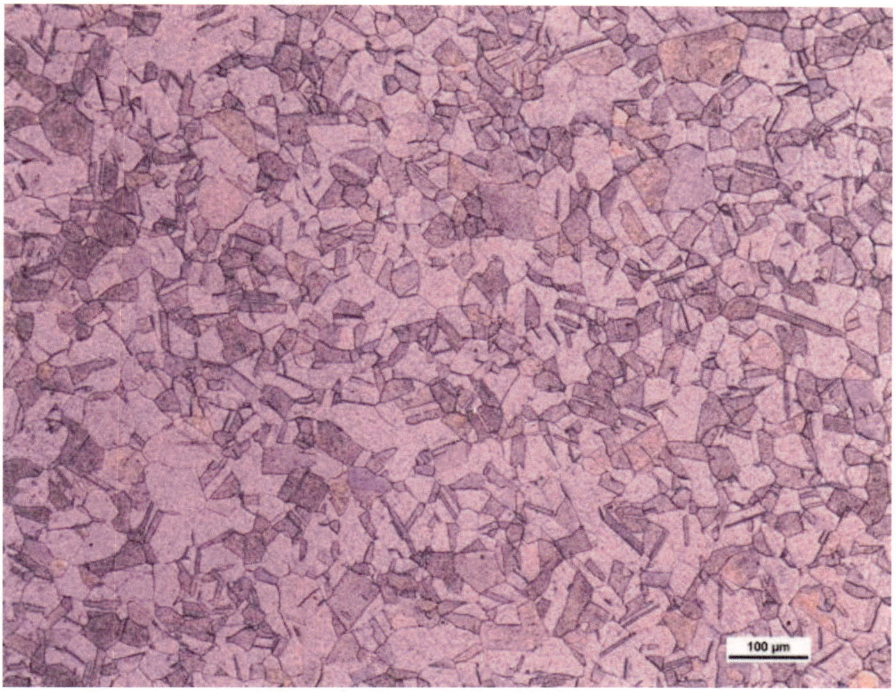

d) the same microstructure of wrought Cu, polarized light

Fig. 4. Pure copper, chemically polished

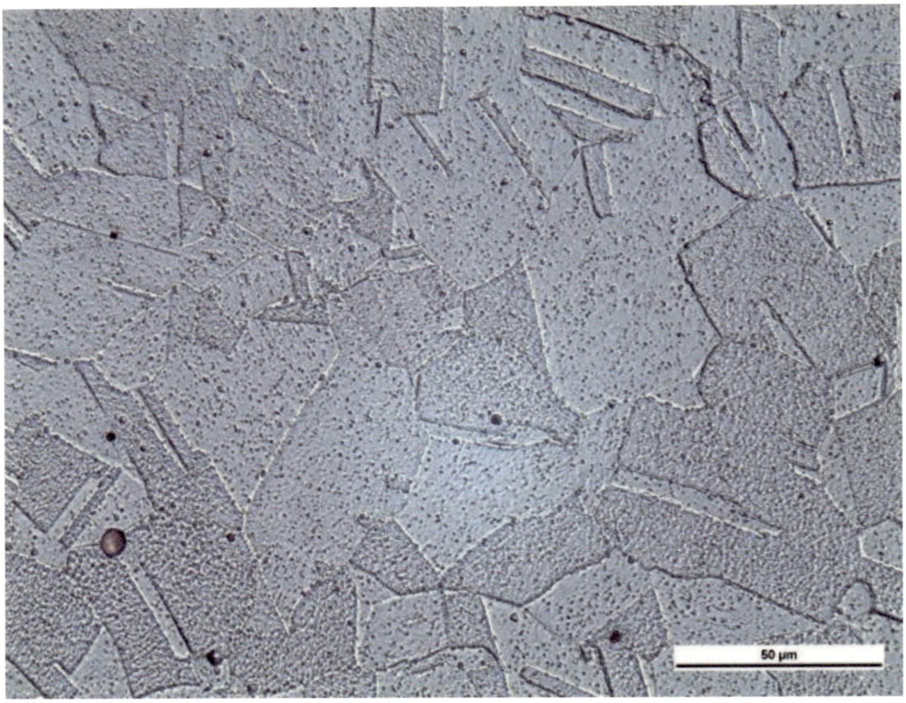

e) detail of wrought Cu grains, white light

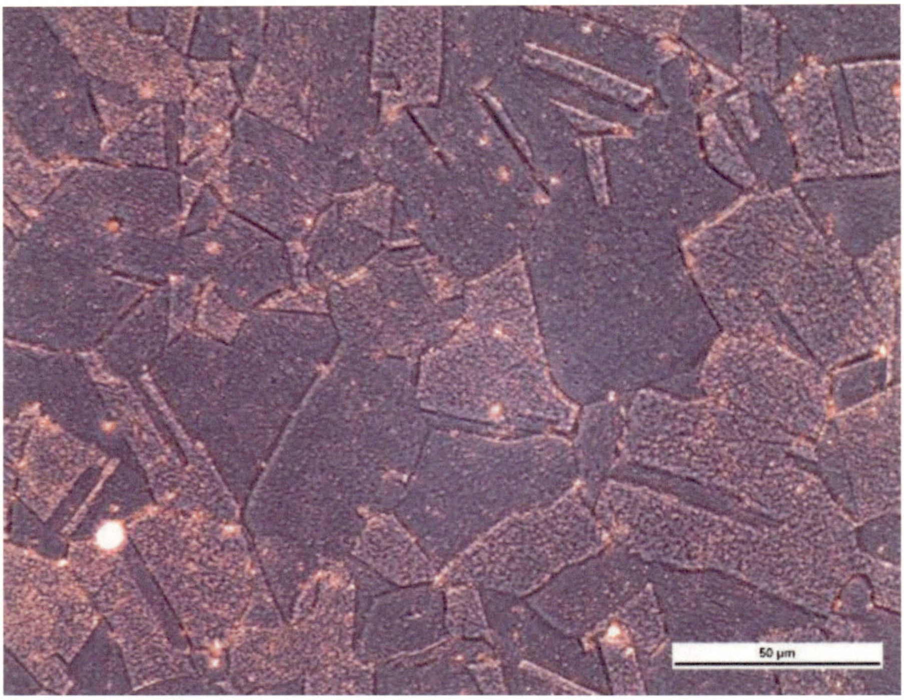

f) detail of wrought Cu grains, polarized light

Fig. 4. Pure copper, chemically polished

The casting characteristics of copper can be improved by the addition of small amounts of elements like beryllium, silicon, nickel, tin, zinc, chromium, and silver. In copper based alloys 14 alloying elements, almost always in the solid solution dissolving area, are used. Most of the industrial alloys are monophasic and they do not show allotropic changes during heating or cooling. For some copper-base alloys precipitation hardening is possible. For the alloys with allotropic recrystallization heat treatment is possible. Single-phase copper alloys are strengthened by cold-working. The FCC copper has excellent ductility and a high strain-hardening coefficient.

Copper and copper based alloys can be divided into 3 groups according to the chemical composition:

- copper and high copper alloys,
- brasses (Cu-Zn + other alloying elements),
- bronzes (Cu + other elements except Zn).

The copper based alloys, according to the application can be divided into two groups, as copper alloys for casting and wrought alloys subsequently. Copper-base alloys are heavier than iron (Skočovský et al., 2006).

Although the yield strength of some alloys is high, their specific strength is typically less than that of aluminum or magnesium alloys. The alloys have better resistance to fatigue, creep, and wear than the lightweight aluminum and magnesium alloys. Many of the alloys have excellent ductility, corrosion resistance, and electrical and thermal conductivity, and they are easily joined or fabricated into useful shapes. The wide varieties of copper-based alloys take advantage of all of the straightening mechanisms on the mechanical properties (Skočovský et al., 2000).

1.2 Copper and copper alloys casting

From the casting point of view, especially the solidification (freezing range) Cu cast alloys can be divided into three groups:

Group I alloys - alloys that have a narrow freezing range, that is, a range of 50 °C between the liquidus and solidus curves. These are the yellow brasses, manganese and aluminum bronzes, nickel bronze, manganese bronze alloys, chromium copper, and copper.

Group II alloys – alloys that have an intermediate freezing range, that is, a freezing range of 50 to 110 °C between the liquidus and the solidus curves. These are the beryllium coppers, silicon bronzes, silicon brass, and copper-nickel alloys.

Group III alloys – alloys that have a wide freezing range. These alloys have a freezing range of well over 110 °C, even up to 170 °C. These are the leaded red and semi-red brasses, tin and leaded tin bronzes, and high leaded tin bronze alloys (R. F. Schmidt, D. G. Schmidt & Sahoo 1998).

According to the cast products quality the Cu based foundry alloys can be classified as high-shrinkage or low-shrinkage alloys. The former class includes the manganese bronzes, aluminum bronzes, silicon bronzes, silicon brasses, and some nickel-silvers. They are more fluid than the low-shrinkage red brasses, more easily poured, and give high-grade castings in the sand, permanent mold, plaster, die, and centrifugal casting processes. With high-shrinkage alloys, careful design is necessary to promote directional solidification, avoid abrupt changes in cross section, avoid notches (by using generous fillets), and properly place gates and risers; all of these design precautions help avoid internal shrinks and cracks (R. F. Schmidt & D. G. Schmidt, 1997).

1.2.1 Casting

To obtain good results from the product quality point of view, the casting processes technological specifications are the most important factor. The lowest possible pouring temperature needed to suit the size and form of the solid metal should be used to encourage as small a grain size as possible, as well as to create a minimum of turbulence of the metal during pouring to prevent the casting defects formation (R. F. Schmidt, D. G. Schmidt & Sahoo, 1988). Liberal use of risers or exothermic compounds ensures adequate molten metal to feed all sections of the casting.

Many types of castings for Cu and its alloys casting, such as sand, shell, investment, permanent mold, chemical sand, centrifugal, and die, can be used (R. F. Schmidt, D. G. Schmidt & Sahoo, 1988). Of course each of them has its advantages and disadvantages. If only a few castings are made and flexibility in casting size and shape is required, the most economical casting method is sand casting. For tin, silicon, aluminum and manganese bronzes, and also yellow brasses, permanent mold casting is best suited. For yellow brasses die casting is well suited, but increasing amounts of permanent mold alloys are also being die cast. Definite limitation for both methods is the casting size, due to the reducing the mold life with larger castings.

All copper alloys can be cast successfully by the centrifugal casting process. Because of their low lead contents, aluminum bronzes, yellow brasses, manganese bronzes, low-nickel bronzes, and silicon brasses and bronzes are best adapted to plaster mold casting. Lead should be held to a minimum for most of these alloys because lead reacts with the calcium sulfate in the plaster, resulting in discoloration of the surface of the casting and increased cleaning and machining costs.

1.2.2 Furnaces

The copper based alloys are melted mainly in Fuel-Fired Furnaces and Electric Induction Furnaces.

From Fuel-Fired Furnaces, oil- and gas-fired furnaces are the most important. However open-flame furnaces are able to melt large amounts of metal quickly; there is a need for operator skill to control the melting atmosphere within the furnace at present this kind of furnaces are not often used. Also, the refractory furnace walls become impregnated with the melting metal causing a contamination problem when switching from one alloy family to another.

When melting leaded red and semi-red brasses, tin and leaded tin bronzes, and high leaded tin bronze alloys, lead and zinc fumes are given off during melting and superheating. These harmful oxides emissions are much lower when the charge is melted in an induction furnace. This is caused by the duration of the melting cycle is only about 25 % of the cycle in a fuel-fired furnace.

The core type, better known as the channel furnace, and the coreless type of electric induction furnaces for Cu and its alloys melting are also used. Because core type furnaces are very efficient and simple to operate with lining life in the millions of pounds poured, they are best suited for continuous production runs in foundries making plumbing alloys of the leaded red and semi-red brasses, tin and leaded tin bronzes, and high leaded tin bronze alloys. They are not recommended for dross-forming alloys; yellow brasses, manganese and aluminum bronzes, nickel bronze, manganese bronze alloys, chromium copper, and copper. The channel furnace is at its best when an inert, floating, cover flux is used and charges of ingot, clean remelt, and clean and dry turnings are added periodically.

Coreless type furnaces have become the most popular melting unit in the Cu alloy foundry industry (R. F. Schmidt, D. G. Schmidt & Sahoo, 1988).

1.2.3 Pure copper and high copper alloys

Commercially pure copper and high copper alloys are very difficult to melt and they are very susceptible to gassing. Chromium copper melting is negatively linked with oxidation loss of chromium. To prevent both oxidation and the pickup of hydrogen from the atmosphere copper and chromium copper should be melted under a floating flux cover. In the case of pure copper crushed graphite should cover the melt. In the case of chromium copper, the cover should be a proprietary flux made for this alloy. It is necessary to deoxidize the melted metal. For this reason the calcium boride or lithium should be plunged into the molten bath when the melted metal reaches 1260 °C. The metal should then be poured without removing the floating cover.

Beryllium coppers can be very toxic and dangerous. This is caused by the beryllium content in cases where beryllium fumes are not captured and exhausted by proper ventilating equipment. To minimize beryllium losses beryllium coppers should be melted quickly under a slightly oxidizing atmosphere. They can be melted and poured successfully at relatively low temperatures. They are very fluid and pour well.

Copper-nickel alloys (90Cu-10Ni and 70Cu-30Ni) must also be carefully melted. Concern is caused by the presence of nickel in high percentages because this raises not only the melting point but also the susceptibility to hydrogen pickup. These alloys are melted in coreless electric induction furnaces, because the melting rate is much faster than it is with a fuel-fired furnace. The metal should be quickly heated to a temperature slightly above the pouring temperature and deoxidized either by the use of one of the proprietary degasifiers used with nickel bronzes or, better yet, by plunging 0.1 % Mg stick to the bottom of the ladle. This has to be done to prevent the of steam-reaction porosity from occurring by the oxygen removing. If the gates and risers are cleaned by shotblasting prior to melting there is a little need to use cover fluxes (R. F. Schmidt, D. G. Schmidt & Sahoo, 1988).

1.2.4 Brasses

Yellow Brasses (containing 23 - 41 % of zinc) as the major alloying element and may contain up to 3 % of lead and up to 1.5 % of tin as additional alloying elements. Due to vaporization these alloys flare, ore lose zinc close to the melting point. The zinc vaporization can be minimized by the addition of aluminum (0.15 to 0.35 %) which also increases the melted metals fluidity. In the case of larger aluminum amount shrinkages take place during freezing; this has to be solved by use of risers. Except for aluminum problems, yellow brass melting is simple and no fluxing is necessary. Zinc lost during the melting should be re-added before pouring.

Silicon brasses have excellent fluidity and can be poured slightly above their freezing range. If overheated, they can pick up hydrogen. In the case of the silicon brasses no cover fluxes are required.

Red brasses and leaded red brasses are copper alloys, containing 2 - 15 % of zinc as the major alloying element and up to 5 % of Sn and up to 8 % of Pb as additional alloying elements. Because of lengthy freezing range in the case of these alloys, chills and risers should be used in conjunction with each other. The best pouring temperature is the lowest one that will pour the molds without having misruns or cold shuts. For good casting, properties retaining these alloys should be melted from charges comprised of ingot and clean, sandfree gates and risers. The melting should be done quickly in a slightly oxidizing atmosphere. When at the proper furnace temperature to allow for handling and cooling to the proper pouring temperature, the crucible is removed or the metal is tapped into a ladle. At this point, a deoxidizer (15 % phosphorus copper) is added. The phosphorus is a reducing agent (deoxidizer) and must be carefully measured so that enough oxygen is removed, yet a small amount remains to improve fluidity. This residual level of phosphorus must be controlled by chemical analysis. Only amount in the range 0.010 and 0.020 % P is accepted, in the case of the larger phosphorus amount internal porosity may occur. Along with phosphorus copper pure zinc should also be added at the point at which skimming and temperature testing take place prior to pouring. This added zinc replaces the zinc lost during melting and superheating. With these alloys, cover fluxes are seldom used. In some foundries in which combustion cannot be properly controlled, oxidizing fluxes are added during melting, followed by final deoxidation by phosphor copper.

Leaded red brasses alloys have practically no feeding range, and it is extremely difficult to get fully sound castings. Leaded red brasses castings contain 1 to 2 % of porosity. Only small castings have porosity below 1 %. Experience has shown that success in achieving good quality castings depends on avoiding slow cooling rates. Directional solidification is the best for relatively large, thick castings and for smaller, thin wall castings, uniform solidification is recommended (R. F. Schmidt, D. G. Schmidt & Sahoo 1998).

1.2.5 Bronzes

Manganese bronzes are carefully compounded yellow brasses with measured quantities of iron, manganese, and aluminum. When the metal is heated at the flare temperature or to the point at which zinc oxide vapor can be detected, it should be removed from the furnace and poured. No fluxing is required with these alloys. The only required addition is zinc, which is caused by its vaporization. The necessary amount is the one which will bring the zinc content back to the original analysis. This varies from very little, if any, when an all-ingot heat is being poured, to several percent if the heat contains a high percentage of remelt.

White manganese bronzes. There are two alloys in this family, both of which are copper-zinc alloys containing a large amount of manganese and, in one case, nickel. They are manganese bronze type alloys, are simple to melt, and can be poured at low temperatures because they are very fluid. The alloy superheating resulting in the zinc vaporization and the chemistry of the alloy is changed. Normally, no fluxes are used with these alloys.

Aluminum bronzes must be melted carefully under an oxidizing atmosphere and heated to the proper furnace temperature. If needed, degasifiers removing the hydrogen and oxygen from the melted metal can be stirred into the melt as the furnace is being tapped. By pouring a blind sprue before tapping and examining the metal after freezing, it is possible to tell whether it shrank or exuded gas. If the sample purged or overflowed the blind sprue during solidification, degassing is necessary. For converting melted metal fluxes, are available, mainly in powder form, and usually fluorides. They are used for the elimination of oxides, which normally form on top of the melt during melting and superheating (R. F. Schmidt, D. G. Schmidt & Sahoo, 1988).

From the freezing range point of view, the manganese and aluminum bronzes are similar to steels. Their freezing ranges are quite narrow, about 40 and 14 °C, respectively. Large castings can be made by the same conventional methods used for steel. The attention has to be given to placement of gates and risers, both those for controlling directional solidification and those for feeding the primary central shrinkage cavity (R. F. Schmidt & D. G. Schmidt, 1997).

Nickel bronzes, also known as nickel silver, are difficult to melt because nickel increases the hydrogen solubility, if the alloy is not melted properly it gases readily. These alloys must be melted under an oxidizing atmosphere and they have to be quickly superheated to the proper furnace temperature to allow for temperature losses during fluxing and handling. After the furnace tapping the proprietary fluxes should be stirred into the metal for the hydrogen and oxygen removing. These fluxes contain manganese, calcium, silicon, magnesium, and phosphorus.

Silicon bronzes are relatively easy to melt and should be poured at the proper pouring temperatures. In the case of overheating the hydrogen, picking up can occur. For degassing, one of the proprietary degasifiers used with aluminum bronze can be successfully used. Normally no cover fluxes are used for these alloys.

Tin and leaded tin bronzes, and **high-leaded tin bronzes**, are treated the same in regard to melting and fluxing. Their treatment is the same as in the case of the red brasses and leaded red brasses, because of the similar freezing range which is long (R. F. Schmidt, D. G. Schmidt & Sahoo, 1988).

Tin bronzes have practically no feeding range, and it is extremely difficult to get fully sound castings. Alloys with such wide freezing ranges form a mushy zone during solidification, resulting in interdendritic shrinkages or microshrinkages. In overcoming this effect, design and riser placement, plus the use of chills, are important and also the solidification speed, for better results the rapid solidification should be ensured. As in the case of leaded red brasses, tin bronzes also have problems with porosity. The castings contain 1 to 2 % of porosity and only small castings have porosity below 1 %. Directional solidification is best for relatively large, thick castings and for smaller, thin wall castings, uniform solidification is recommended. Sections up to 25 mm in thickness are routinely cast. Sections up to 50 mm thick can be cast, but only with difficulty and under carefully controlled conditions (R. F. Schmidt & D. G. Schmidt, 1997).

2. Copper based alloys

2.1 Brasses

Brasses are copper based alloys where zinc is the main alloying element. Besides zinc, also some amount of impurities and very often some other alloying elements are present in the alloys. The used alloying elements can improve some properties, depending on their application. Due to the treatment, brasses can be divided into two groups: wrought brasses and cast brasses. One special group of brasses is brazing solder.

The binary diagram of the Cu-Zn system is quite difficult, Fig. 5. For the technical praxis only the area between the 0 to 50 % of Zn concentration is important. Alloys with higher Zn concentration have not convenient properties, as they are brittle.

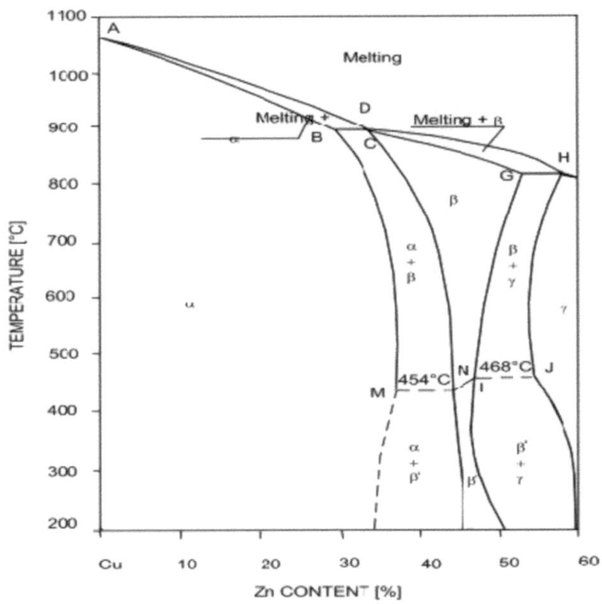

Fig. 5. Binary diagram copper-zinc

In the liquid state copper and zinc are absolutely soluble. In the solid state solubility is limited; copper has face centered cubic lattice and zinc has the hexagonal close-packed lattice (HCP). During solidification the α, β, γ, δ, ε and η phases are released. Most of the phases are intermediary phases characterized by the relation of valence electrons and amount of atoms. Primary solid solution α (Zn in Cu) has the same crystal lattice as pure copper and dissolves maximum 39 % of Zn at temperature 450 °C. Decreasing the temperature, also the zinc solubility in the α solution decreases ~ 33 %. In the alloys with higher content of Zn concentration the supersaturated α solid solution is retaining at room temperature. This is why alloys up to 39 % Zn concentration have homogenous structure consisting from the α solid solution crystals. After the forming and annealing the brass structure consists of the α solid solution polyhedral grains with annealed twins (Skočovský et al., 2000, 2006).

Alloys with Zn concentration from 32 % (point B, Fig. 5) to 36 % (point C) crystallize according to the peritectic reaction at temperature 902 °C. During this reaction the β phase is created by the reaction of formed primary α crystals and the liquid alloy. By decreasing the temperature in the solid state the ratio of both crystals is changed. This is caused by solubility changes and the resulting brass structure is created only by the α solid solution crystals. In the case of alloys with the concentration of Zn from 36 to 56 % the β phase exists after the solidification. β phase has a body centered cubic (BCC) lattice. The atoms of Cu and Zn are randomly distributed in the lattice. β phase has good ductility. By the next temperature decreasing the random β phase is changed to the ordered hard and brittle β´ phase; temperatures from 454 to 468 °C. The 39 to 45 % Zn containing brass resulting structure is heterogeneous, consisting of the α solid solution crystals and β´ phase crystals.

Only the alloys containing less than 45 % of Zn in the technical praxis, apart from small exceptions, are used. Brasses with less than 40 % Zn form single-phase solid solution of zinc in copper. The mechanical properties, even elongation, increase with increasing zinc content. These alloys can be cold-formed into rather complicated yet corrosion-resistant components. Brasses with higher Zn concentration, created by the β´ phases, or β´+γ phase, are excessively brittle. Mechanical properties of the brasses used in the technical praxis are shown in the Fig. 6 (Skočovský et al., 2000).

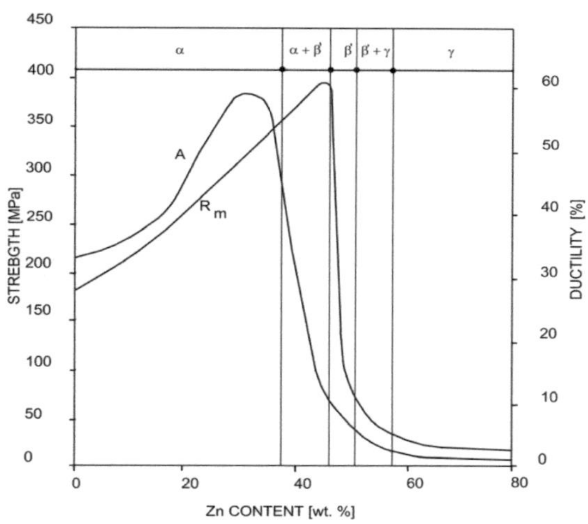

Fig. 6. Influence of the Zn content to the brass mechanical properties

Brasses used in technical praxis consist also of some other elements; impurities and alloying elements besides Zn, whose influence is the same as in the case of pure copper. Bismuth and sulphur, for example, decreases the ability of the metal to hot forming. Lead has a similar influence. On the other hand, lead improves the materials workability. For this reason, lead is used for heterogeneous brasses in an amount from 1 to 3 %. Homogenous brasses, for improving strength, contain tin, aluminum, silicon and nickel. Silicon improves the materials resistance against corrosion and nickel improves the materials ductility.

Brasses heat treatment. Recrystallization annealing is brasses basic heat treatment process. Combination of recrystallization annealing and forming allows to change the materials grain structure and to influence the hardening state reached by the plastic deformation. The lower limit of the recrystallizing temperatures for binary Cu-Zn alloys is ~ 425 °C and the upper limit is limited by the amount of Zn in the alloy.

Stress-relief annealing is the second possible heat treatment used for brasses. This heat treatment process, at temperatures from 250 - 300 °C, decreases the danger of corrosion cracking (Skočovský et al., 2000).

2.1.1 Wrought brasses

Wrought brasses are supplied in the form of metal sheets, strips, bars, tubes, wires, etc. in the soft (annealed) state, or in the state (medium, hard state) after cold forming (Skočovský et al., 2000, 2006).

Tombacs are brasses containing more than 80 % of copper. They are similar to the pure copper by theirs chemical and physical properties, but they have better mechanical properties. Tombac with higher copper content is used for coins, memorial tablets, medals, etc. (after production the products are gold-plated before distribution). Tombacs with medium or lower copper content are light yellow colored, close to the color of gold. Because of this, brass films are used like the gold substituent in the case of decorative, artistic, fake jewelry, architecture, armatures, in electrical engineering, for manometers, flexible metal tubes, sieves etc. plating. Tombac with 80 % Zn has lower chemical resistance due to the higher Zn content, and so its usage (same like for other Tombacs) is dependent on the working conditions.

Deep-drawing brasses contain around 70 % of Cu (Fig. 7); for securing high ductility the impurities elements content has to be low. Cu-Zn alloys with the 32 % Zn content have the highest ductility at high strength which is why those alloys are used for deep-drawing. They are used, for example, for ball cartridges, musical equipment production and in the food industry (Skočovský et al., 2000).

Brass with higher Zn content (37 % of Zn) is quite cheap because of its lower Cu content. It is a heterogeneous alloy with small β´ phase content, with lower ductility and good ability for cold forming, Fig. 8. It is used for different not very hard loaded products; for example wiring material in electrical engineering, automotive coolers, etc. For improving the materials workability a small amount of lead is added to brasses with higher Zn content (Skočovský et al., 2000).

Brass with 40 % of Zn is heterogeneous alloy with (α+β´) microstructure. Compared to other brasses it has higher strength properties, but lower ductility and cold forming ability. This is caused by the β´ presence. It is suitable for forging and die pressing at higher temperatures (700 - 800 °C). This kind of brass is used in architecture, for different ship forging products, for tubes and welding electrodes. For this brass type, with 60 % of Cu, the tendency to dezincification corrosion and to corrosion cracking is typical. The crack tendency increases with the increasing Zn content and it is largest at 40 % Zn content. In the case of zinc content below 15 % this tendency is not present. Brasses alloying elements do not improve the crack tendency. Some of the used alloying elements can decrease this disadvantage; Mg, Sn, Be, Mn. Grain refinement has the same influence (Skočovský et al., 2000).

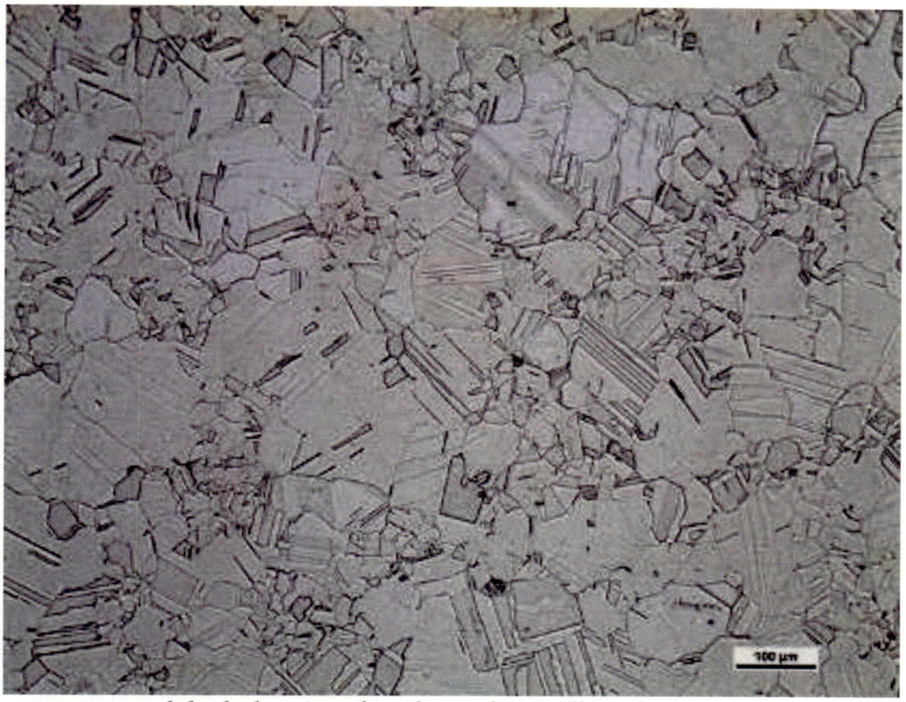

a) microstructure, polyhedral grains of α phase, chemically etched

b) size and orientation of grains, polarized light, etched $K_2Cr_2O_7$

Fig. 7. Deep-drawing brass with 70 % Cu

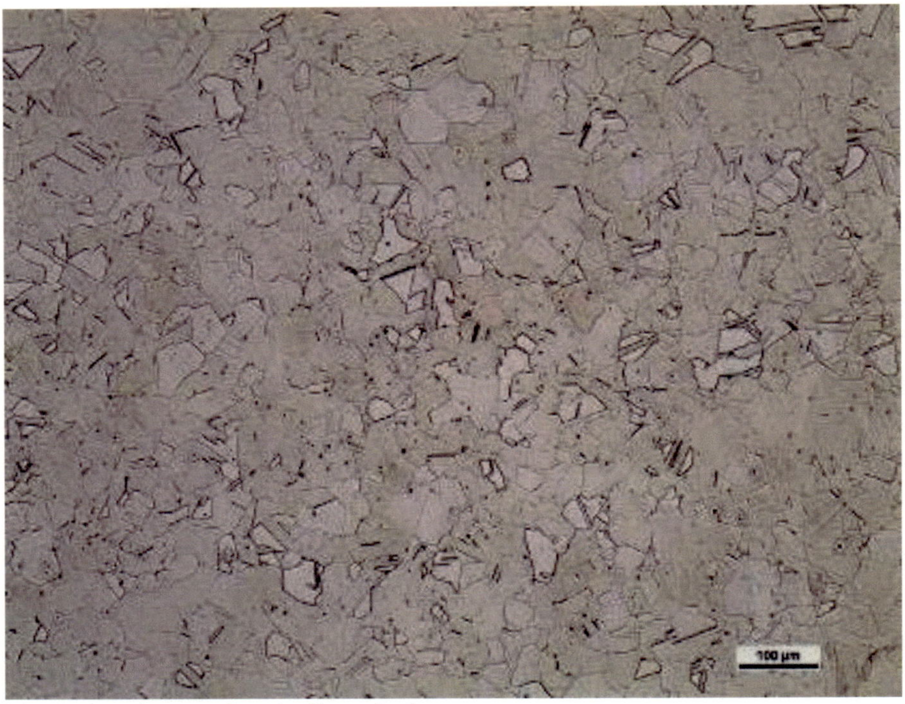

a) (α+β´) microstructure, chemically etched

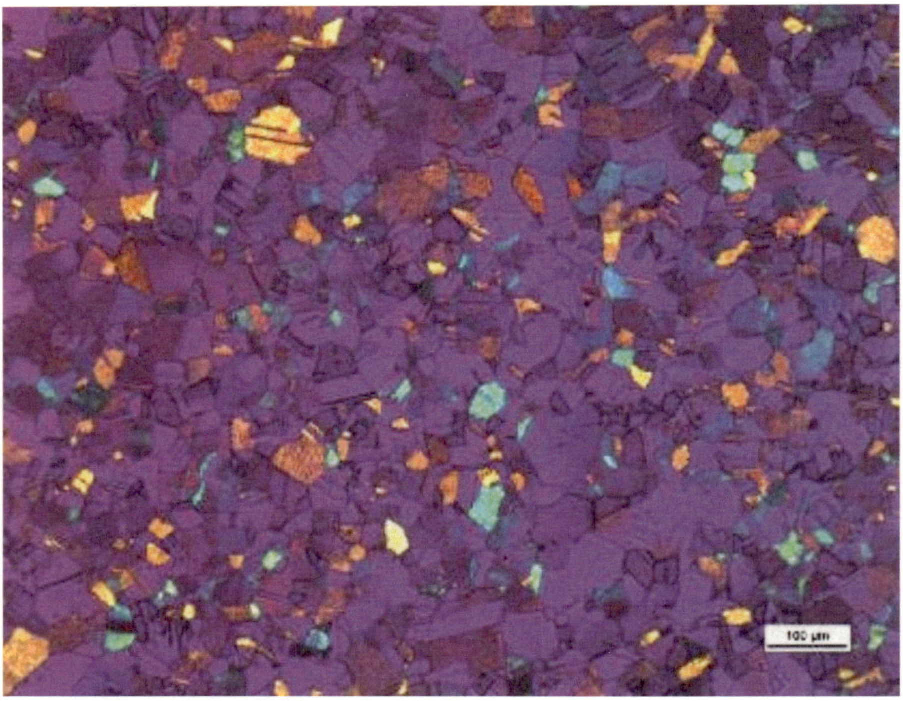

b) α phase is violet, polarized light, etched $K_2Cr_2O_7$

Fig. 8. Brass with 37 % of Zn

Brass CuZn40Mn3Fe1 is an alloy with a two phase microstructure (Fig. 9); α phase is light (in black and white color) or blue (with polarized light) and β´ phase is dark or a different color depending on the grains orientation. The microstructure shows polyhedral grains with different size and color of β´ phase (Fig. 9a). A direction of α phase is different according to the orientation of grains, as is shown in a detail of the microstructure (Fig. 9b).

Leaded brass. Brasses in this group contain ~ 60 % of copper and from 1 to 3 % of lead which improve the materials machinability (Skočovský et al., 2000). The microstructure is constituted by two phases, α and β´, where α phase is lighter than the second phase β´, (Fig. 10).

Lead is not soluble in copper and so its influence is the same as in the case of leaded steels. Lead should be finely dispersed as isolated particles in the final brass structure (Fig. 10). Smaller carburetor parts and seamless tubes are produced from leaded brasses.

Brasses with 58 or 59 % of Cu and lead are suitable for metal sheets, strips, bands, bars production and different shaped tubes. At higher temperatures it is possible apply forging and pressing to these brasses. They are difficult to cold form, but it is possible to strike them. They are used for different stroked product for watchmakers and small machines, especially in electrical engineering production. Leaded brass with 58 % of Cu and is used for screw production and for other in mass scale turned products (Skočovský et al., 2000).

The CuZn43Mn4Pb3Fe brass with higher content of Zn (43 %) has only one phase β´ microstructure (Fig. 11). This phase is constituted by polyhedral grains (different color according to the grains orientation) with very small black globular particles of the phase with high content of iron, and light colored regularly distributed small Pb particles.

2.1.2 Cast brasses

Cast brasses are heterogeneous alloys (α+β´) containing from 58 to 63 % of copper. For improving machinability they very often also contain lead (1 - 3 %). Cast brasses have good feeding, low tendency to chemical unmixing, but high shrinkage. Because of their structure cast brasses have lower mechanical properties compared to the mechanical properties of wrought brasses. These brasses are used mainly for low stressed castings; pumps parts, gas and water armatures, ironworks for furniture and building, etc. (Skočovský et al., 2000, 2006).

2.1.3 Special brasses

This alloy family consists of copper, zinc and one or more elements in addition (aluminum, tin, nickel, manganese, iron, silicon). The name of the alloy is usually formed according to the additional element (for example silicon brass is the Cu-Zn-Si alloy). Other elements addition improves the materials mechanical properties, corrosion resistance, castability and workability increasing. The change in properties is dependant on the element type and on its influence on the materials structure.

According to production technology special brasses can be divided into two groups: wrought and cast special brasses (Skočovský et al., 2000, 2006).

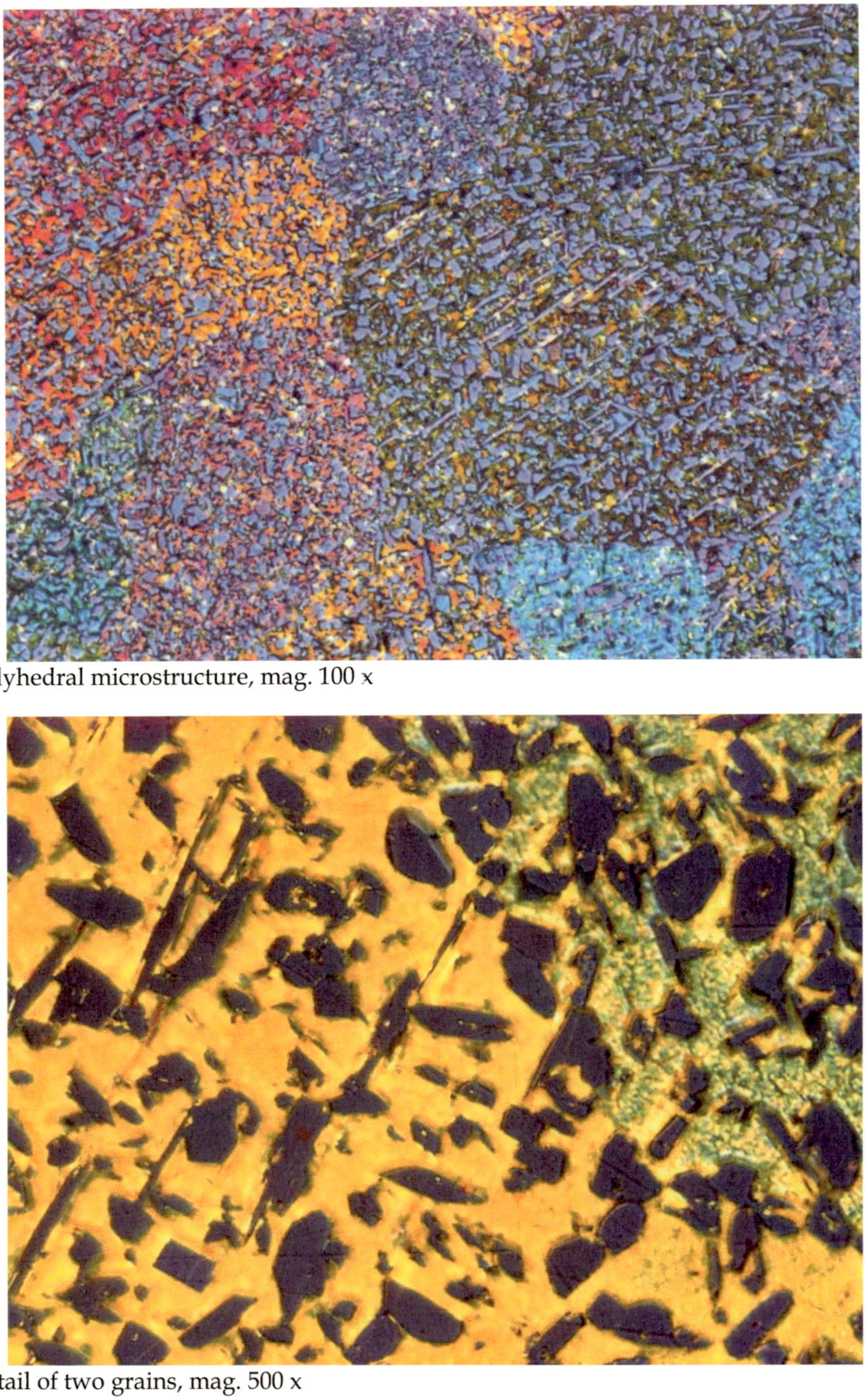

a) polyhedral microstructure, mag. 100 x

b) detail of two grains, mag. 500 x

Fig. 9. Brass CuZn40Mn3Fe1, polarized light, etched $K_2Cr_2O_7$

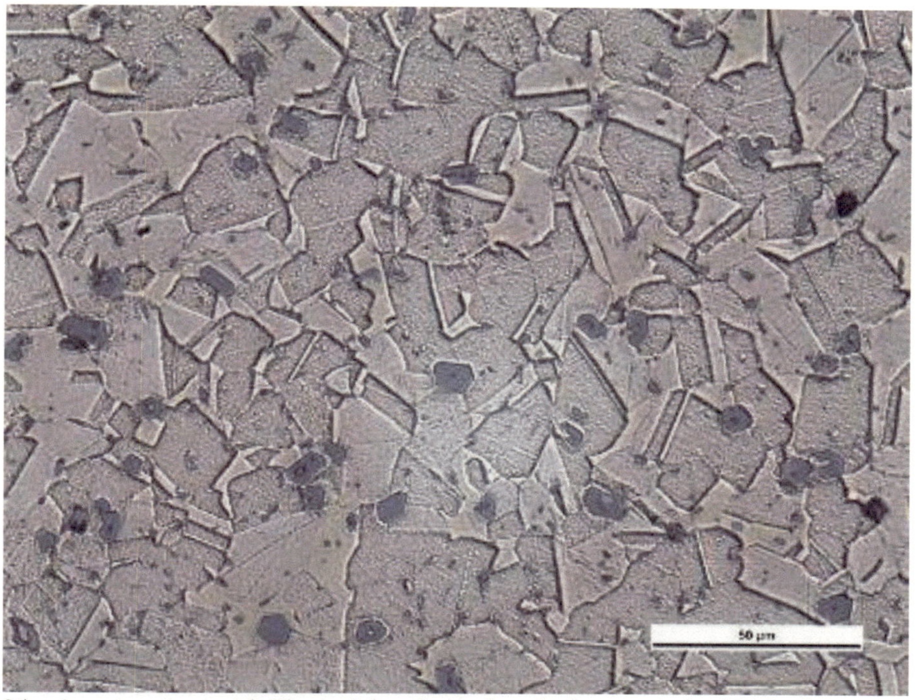

a) (α+β´) microstructure with Pb particles showed as dark gray, etched $K_2Cr_2O_7$

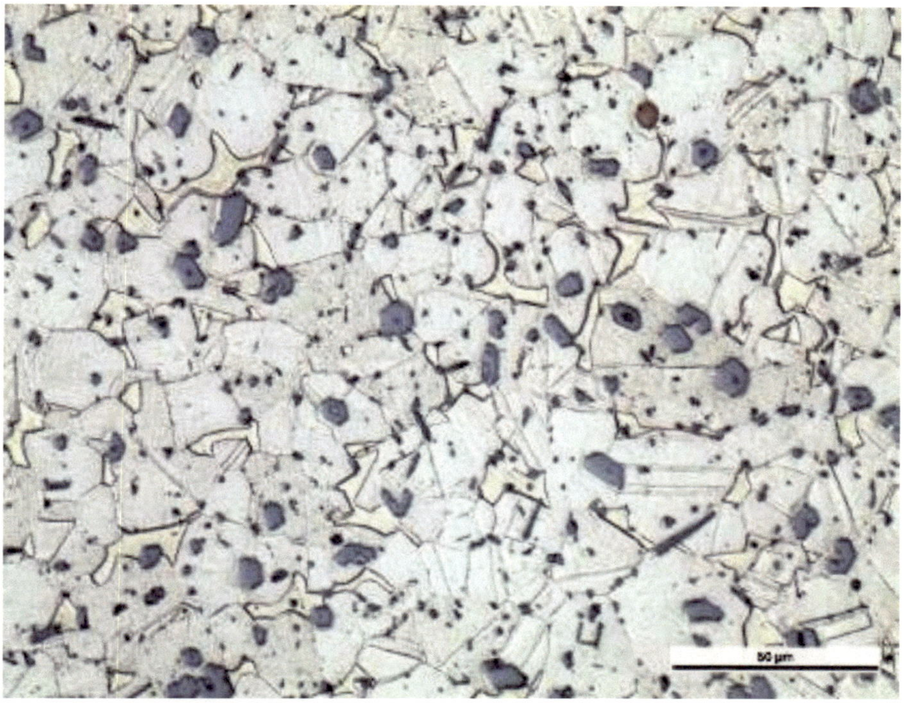

b) the same microstructure after chemical etching

Fig. 10. Leaded brass, 42 % of Zn and 2 % of Pb

a) polyhedral grains of β´phase, polarized light, mag. 100 x

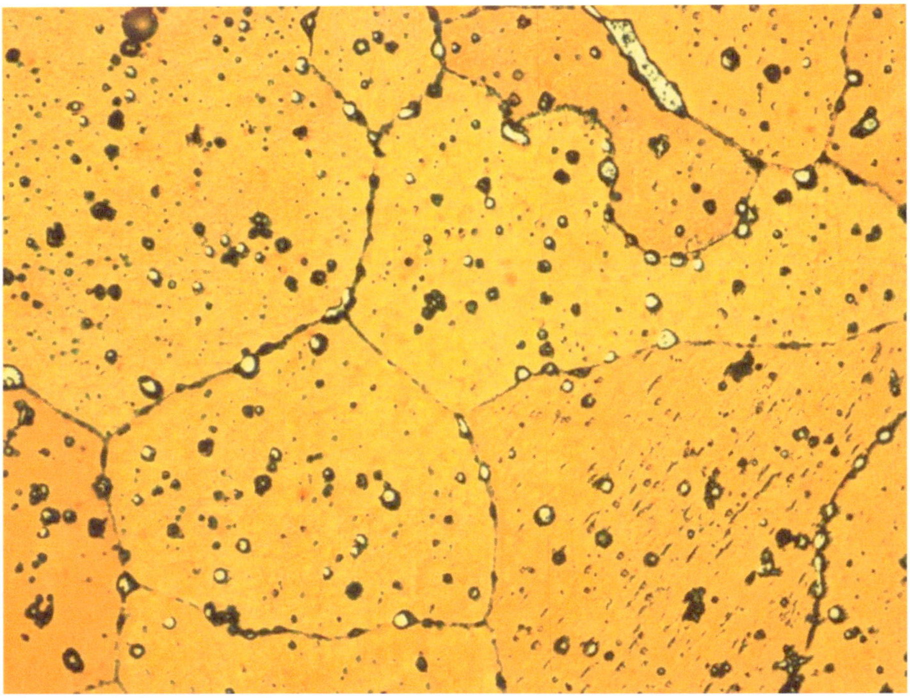

b) detail, white light, mag. 500 x

Fig. 11. Brass CuZn43Mn4Pb3Fe, etched $K_2Cr_2O_7$

Aluminum brasses. Aluminum brasses contain from 69 to 79 % of copper; the aluminum additive content, in the case of wrought alloys, is below 3 to 3.5 % to keep the structure homogeneous. As well as this aluminum content, the structure is also formed by new phases as are $\beta + \gamma$ phases, which improve the materials hardness and strength, but decrease its ductility. Aluminum brass containing 70 % of Cu and 0.6 to 1.6 % Al, with Sn and Mn addition, is very corrosion resistant and is used for condenser tubes production. Al brass with higher content of Cu (77 %) and with Al from 1.7 to 2.5 %, whose application is the same as that of the previous brass, its corrosion resistance against the see water is higher because of the larger Al content.

The structure of cast aluminum brasses is heterogeneous. The copper content is in this case lower and the aluminum content is higher (below 7 %), which ensures good corrosion resistance of the material in sea water. They are used for very hard loaded cast parts; armatures, screw-cutting wheel, bearings and bearings cases.

Manganese brasses. Wrought manganese brasses contain from 3 to 4 % manganese and cast manganese brasses contain from 4 to 5 % manganese. This alloying family has high strength properties, and corrosion resistance. They are used usually in the heterogeneous structure. Wrought manganese brasses with 58 % of Cu or 57 % of Cu with addition of Al have quite good strength (in medium-hard state 400 to 500 MPa) at large toughness and corrosion resistance. They are used for armatures, valve seating, high-pressured tubes, etc. Mn brass with 58 % of Cu is also used for decorations (product surface is layered during hot oxidation process by attractive, durable brown verdigris). Cast manganese brasses have larger manganese and iron content and they are used for very hard loaded castings; weapons parts, screw propellers, turbine blades and armatures for the highest pressures.

Tin brasses. Tin brasses are mostly heterogeneous alloys containing tin below 2.5 %. Tin has a positive influence on mechanical properties and corrosion resistance. Some Sn alloys with 60 % of Cu have good acoustical properties and so they are used for the musical/sound instruments production. Sn brass alloy with 62 % of Cu is used for the strips, metal sheets and profiles used in the ships or boat constructions.

Silicon brasses. Silicon brasses contain maximum 5 % of Si at large copper content, from 79 to 81 %. As in the other brasses case, silicon brasses can be wrought (max. 4 % Si) or cast. They have very good corrosion resistance and mechanical properties also at low temperatures (- 183 °C). Over 230 °C their creeping is extensive, at a low stress level and at a temperature larger than 290 °C silicon brasses are also brittle. Lead addition, (3 to 3.5 %), positively affects the materials wear properties and so these alloys are suitable for bearings and bearings cases casting (Cu 80 %, Si 3 %, Pb 3 %). Silicon brasses are used in boats, locomotives and railway cars production. Silicon cast alloys with good feeding properties are used for armatures, bearings cases, geared pinions and cogwheels production.

Nickel brasses. Nickel brasses contain from 8 to 20 % nickel, which is absolutely soluble in homogeneous brasses α and nickel enlarges α area. Homogeneous alloys are good cold formed and are suitable for deep-drawing. Heterogeneous alloys (α+β) are good for hot forming. Nickel brasses have good mechanical properties, corrosion resistance and are easily polishable. One of the oldest alloys are alloys of 60 % of copper and from 14 to 18 % of nickel content; they were used for decorative and useful objects. These alloys have many different commercial names, for example "pakfong", "alpaca", "argentan" etc. they are used in the building industry, precise mechanics, and medical equipment production and for stressed springs (high modulus of elasticity).

Brazing solders. Brazing solders are either basic or special brasses with melting temperature higher than 600 °C. They are used for soldering of metals and alloys with higher melting temperatures like copper and its alloys, steels, cast irons etc. In the case of binary alloys Cu-Zn with other element addition (Ag or Ni), the solders are marked like silver or nickel solders. Brasses solders with Cu content from 42 to 45 % have the melting temperature 840 to 880 °C and they are used for brasses soldering. Silver solders contain from 30 to 50 % of copper, from 25 to 52 % of Zn and from 4 to 45 % of Ag. Lower the Ag content; lower the melting temperature (to 720 °C). Silver solders have good feeding and give strong soldering joints. Nickel solders (38 % Cu, 50 % Zn and 12 % Ni) have a melting temperature of around 900 °C. They are used for steels and nickel alloys soldering.

Solder with a very low Ag content (0.2 to 0.4 %) with upper melting temperature 900 °C has good electric conductivity and for this reason is used in electrical engineering. Solder with an upper melting temperature of 850 °C (60 % Cu and low content of Si, Sn) has high strength and it is suitable for steels, grey cast iron, copper and brasses soldering (Skočovský et al., 2000, 2006).

2.2 Bronzes

Bronzes are copper based alloys with other alloying elements except zinc. The name of bronzes is defined according to the main alloying element; tin bronzes, aluminum bronzes, etc. (Skočovský et al., 2000, 2006).

2.2.1 Tin bronzes

Tin bronzes are alloys of copper and tin, with a minimal Cu-Sn content 99.3 %. Equilibrium diagram of Cu-Sn is one of the very difficult binary diagrams and in some areas (especially between 20 to 40 % of Sn) it is not specified till now. For the technical praxis only alloys containing less than 20 % of Sn are important. Tin bronzes with higher Sn content are very brittle due to the intermetallic phases' presence. Cu and Sn are absolutely soluble in the liquid state. In the solid state the Cu and Sn solubility is limited.

Normally, the technical alloys crystallize differently as compared to the equilibrium diagram. Until 5 % of Sn, the alloys are homogenous and consist only of the α solid solution (solid solution of Sn in Cu) with face centered cubic lattice. In the cast state the alloy structure is dendritic and in the wrought and annealed state the structure is created by the regular polyhedral grains. The resulting structure of alloys with larger Sn content (from 5 to 20 %) is created by α solid solution crystals and eutectic ($\alpha + \delta$). δ phase is an electron compound $Cu_{31}Sn_8$ ($e/a = 21/13$) with cubic lattice. δ phase is brittle phase, which has negative influence on the ductility and also decreases the materials strength in case of higher Sn content (above 20 %). Even though the solubility in the case of technical alloys decreases, the ε phase (Cu_3Sn with hexagonal lattice; $e/a = 7/4$) is not created. The ε phases do not occur because the diffusion ability of Sn atoms below 350 °C is low. ε phase also does not occur at normal temperature with higher Sn content in bronze.

Tin addition has a similar influence on bronzes properties as zinc addition in the case of brasses. For the forming, bronzes with around 9 % of Sn are used (it is possible to heat those alloys to single-phase state above 5 % Sn). Tin bronzes are used when bronzes are not sufficient in strength and corrosion resistance points of view. For casting, bronzes with higher Sn content are used; up to 20 % of Sn. Cast bronzes are used more often than wrought bronzes. Tin bronzes castings have good strength and toughness, high corrosion resistance and also good wear properties (the wear resistance is given by the heterogeneous structure ($\alpha + (\alpha + \delta)$). Tin bronzes have small shrinkages during the solidification (1 %) but they have worst feeding properties and larger tendency to the creation of microshrinkages.

Wrought tin bronzes

Bronze CuSn1 contains from 0.8 to 2 % of Sn. In the soft state this bronze has tensile strength 250 MPa and 33 % ductility. It has good corrosion resistance and electric conductivity; it is used in electrical engineering. Bronze CuSn3 with 2.5 to 4 % of Sn has in its soft state tensile strength 280 MPa and ductility 40 %. It is used for the chemical industry and electric engineering equipment production. Bronze CuSn6 with tensile strength 350 MPa and ductility 40 % (in soft state) is used for applications where β', a higher corrosion resistance is required for good strength properties and ductility; for example corrosion environment springs. CuSn8 bronze has, from all wrought tin bronzes the highest strength (380 MPa) and ductility (40 %). It is suitable for bearing sleeves production and in the hard state also for springs which are resistant to fatigue corrosion.

Cast tin bronzes

Bronze CuSn1 with low Sn content has sufficient electric conductivity and so it is used for the castings used in electric engineering. CuSn5 and CuSn10 bronzes have tensile strength 180 and 220 MPa, ductility 15 % and they have good corrosion resistance. They are used for the stressed parts of turbines, compressors, for armatures and for pumps runners' production. Bronze CuSn12 is used for parts used to large mechanical stress and wear frictional loading; spiral gears, gear rims. CuSn10 and CuSn12 bronzes are used in the same way as bearing bronzes. High Sn content (14 to 16 %) bronzes usage have been, because of their expense, replaced by lower Sn containing bronzes, around 6 %, with good sliding properties (Skočovský et al., 2000, 2006).

2.2.2 Leaded tin bronzes and leaded bronzes

Leaded tin bronzes and leaded bronzes are copper alloys where the Sn content is partially or absolutely replaced by Pb. The Pb addition to copper, improves the alloys sliding properties without the negative influence on their heat conductivity. Cu-Pb system is characteristic by only partial solubility in a liquid state and absolute insolubility in a solid state. The resulting structure, after solidification, consists of copper and lead crystals. At a high cooling rate both the alloy components are uniformly distributed and the alloys have very good sliding properties. Leaded bronzes are suitable for steel friction bearing shells casting. They endure high specific presses, quite high circumferential speeds and it is possible to use them at elevated temperatures (around 300 °C).

Two types of bearing bronzes are produced. Bronzes with lower Pb content (from 10 to 20 %) and Sn addition (from 5 to 10 %) and also high-leaded bronzes (from 25 to 30 %) without tin. At present, specially leaded bronzes CuSn10Pb and CuSn10Pb10 like bearing bronzes are used (Skočovský et al., 2000, 2006). Lead (additive from 4 to 25 %) improves bearing sliding properties, and tin (from 4 to 10 %) improves strength and fatigue resistance. These alloys are used especially for bearings in dusty and corrosive environments. Second group binary alloys have lower strength and hardness and they are used for steel shells coatings. With small additions of Mn, Ni, Sn and Zn (in total 2 %) it is possible to refine the structure and to decrease the materials tendency to exsolution. These are very often used for steel shells coated by thin leaded bronze layer for the main and the piston rod bearings of internal-combustion engine (Skočovský et al., 2000).

2.2.3 Aluminum bronzes

Aluminum bronzes are alloys of copper, where aluminum is the main alloying element. For the technical praxis alloys with Al content below 12 % are important. Equilibrium diagram of Cu-Al is complicated and it is similar to Cu-Sn equilibrium diagram. One part of Cu-Al system for alloys containing up to 14 % of Al is shown on Fig. 12.

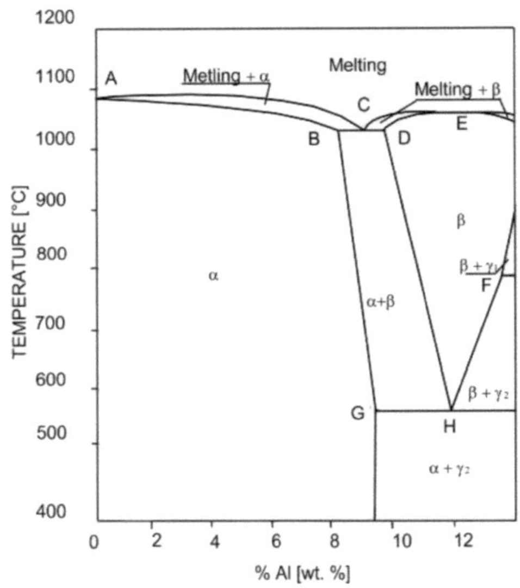

Fig. 12. Part of Cu-Al system equilibrium diagram

The solubility of Al in copper is maximum 7.3 % but it grows with temperature increasing to 9.4 % Al. Homogeneous alloys structure is created by α solid solution crystals (substituted solid solution of Al in Cu) with body centered cubic lattice with similar properties as has the α solid solution in brasses. It is relatively soft and plastic phase. In the real alloys the absolutely equilibrium state does not occur. In the case of Al content close to the solubility limit some portion of β phase in the structure will occur. The upper limit of Al in α homogeneous structure alloys is dependent on the cooling rate and it is in the range of 7.5 to 8.5 % of Al.

Alloys with Al content in the range of 7.3 to 9.4 % solidify at eutectic reaction (α + β) and close to the eutectic line they contain primary released phase α or β and eutectic. (After the changed at lower temperatures the eutectic disappears and so its influence in the structure cannot be proven.) By decreasing the temperature the composition, of α and β crystals changes according to the time of solubility change. β phase is a disordered solid solution of electron compound Cu_3Al (e/a = 3/2) with face centered cubic lattice. It is a hard and brittle phase. β phase, from which the α solid solution is created at lower temperatures, is precipitated from liquid metal at the Al content from 9.5 to 12 % alloys during the crystallization process. During the slow cooling rate the β phase is transformed at eutectoid temperature 565 °C to the lamellar eutectoid ($\alpha + \gamma_2$). For this reason the eutectoid reaction of β phase is sometimes called "pearlitic transformation".

Phase γ_2 is solid solution of hard and brittle electron compound Cu_9Al_4 with complicated cubic lattice. After the recrystallization in solid state the slowly cooled alloys with Al content from 9.4 to 12 % are heterogeneous. Their structure is created with α solid solution crystals and eutectoid ($\alpha + \gamma_2$).

Because of the possibility to improve the mechanical properties by heat treatment heterogeneous alloys are used more often than homogeneous alloys. From 10 to 12 % Al content alloys can be heat treated with a similarly process as in the case of steels. The martensitic transformation can be reached in the case when the eutectoid transformation is limited by fast alloys cooling rate from the temperatures in the β or ($\alpha + \beta$) areas (Fig. 12). After this process the microstructure with very fine and hard needles β_1 phase with a body centered cubic lattice will be reached. By the β phase undercooling below the martensitic transformation temperature M_s, a needle-like martensitic supersaturated disordered solid solution β´ phase with body centered cubic lattice is created.

Due to the chemical composition aluminum bronzes can be divided into two basic groups:

- elementary (binary) alloys; i.e. Cu-Al alloys without any other alloying elements,
- complex (multicomponent) alloys; besides the Al these alloys contain also other alloying elements like Fe, Ni, Mg whose content does not exceeds 6 %.

Iron is frequently an aluminum bronzes alloying element. It is dissolved in α phase till 2 % and it improves its strength properties. With Al it creates $FeAl_3$ intermetallic phase which causes the structure fining.

Manganese is added to the multicomponent alloys because it has deoxidizing effect in the melted metal. It is dissolved in α phase, up to 12 % of Mn content, and it has an effect similar to iron.

Nickel is the most frequent alloying element in aluminum bronzes. It has positive influence on the corrosion resistance in aggressive water solutions and in sea water. Up to around 5 % nickel is soluble in α phase. Nickel with aluminum creates Ni_3Al intermetallic phase which has a precipitate hardening effect.

Homogeneous aluminum bronzes are tough and are suitable for cold and also hot forming. Heterogeneous alloys are stronger, harder, but they have lower cold forming properties compared to the homogeneous alloys. They are suitable for hot forming and have good cast properties. Aluminum bronzes are distinguished by good strength, even at elevated temperatures, and also very good corrosion resistance and wear resistance. Aluminum bronzes are used in the chemical and food industry for stressed components production.

These alloys are used in the mechanical engineering for much stressed gearwheels and worm wheels, armatures working at elevated temperatures etc. Production due to the treatment the aluminum bronzes are divided into two groups; cast and wrought aluminum bronzes.

Aluminum bronzes with Al content from 4.5 to 11 % are used for forming elementary or complex. Al content from 7.5 to 12 % are used for casting only complex aluminum bronzes

CuAl15 bronze is used for cold forming. It is supplied in the form of sheets, strips, bars, wires and pipes. In the soft state this alloy can reach the tensile strength 380 MPa, ductility 40 % and hardness 70 to 110 HB. It is used in the boats building, chemical, food and paper making industry.

Complex aluminum bronzes are normally used for hot forming. CuAl9Mn2 is used for the armatures (bellow 250 °C) production. CuAl9Fe3 is used for the bearings shells, valve seats production, etc. CuAl10Fe3Mn1.5 alloy has heightened hardness and strength; it is suitable for shells and bearings production; it is replacing leaded bronzes up to temperature 500 °C, sometimes also till 600 °C, the CuAl10Fe4Ni4 where Ni is replacing Mn is used. Nickel positively affects materials mechanical and corrosion properties. After the heat treatment the alloy has the tensile strength of 836 MPa and ductility 13.4 %. In the sea water corrosion environment this bronze reached better results compared to chrome-nickel corrosion steels. It is resistant against cavitational corrosion and stress corrosion. CuAl10Fe4Ni4 is used for castings, also used for water turbines and pumps construction, for valve seats, exhaust valves and other components working at elevated temperatures and also in the chemical industry. Besides CuAl19Ni5Fe1Mn1 the nickel alloy consists also a higher content of manganese. It is suitable for cars worm wheels, compressing rings of friction bearings for high pressures etc. (Skočovský et al., 2000, 2006).

2.2.4 Silicon bronzes

The silicon content in this type of alloys is in the range from 0.9 to 3.5 %. The Si content should not exceed 1 % when higher electric conductivity is required. Silicon bronzes more often in the form of complex alloys Cu-Si-Ni-Mn-Zn-Pb are produced; binary alloys Cu-Si only rarely are used. Manganese is dissolved in the solid solution; improving strength, hardness and corrosion properties. Zinc improves the casting properties and mechanical properties, as same as Mn. Nickel is dissolved in the solid solution but it also creates Ni_2Si phase with silicon, which has a positive influence on the materials warm strength properties. Lead addition secure sliding properties.

Silicon bronzes have good cold and hot forming properties and are also used for castings production. They are resistant against sulphuric acid, hydrochloric acid and against some alkalis. Because of their good mechanical, chemical and wear properties, silicon bronzes are used for tin bronzes replacing; they outperform tin bronzes with higher strength and higher working temperatures interval. Formed CuSi3Mn alloy has in the soft state tensile strength 380 MPa and ductility 40 %. It is used for bars, wires, sheets, strips, forgings and stampings production. Casting alloys have normally higher alloying elements content and Si content reaches 5 % very often (Skočovský et al., 2000).

2.2.5 Beryllium bronzes

Beryllium is in copper limitedly soluble (max. 2.7 %) and in the solid state the solubility decreases (0.2 % at room temperature). The binary alloys with low beryllium content (0.25 to 0.7 %) have good electric conductivity, but lower mechanical properties, they are used rarely. More often alloys with higher Be content and other alloying elements as Ni, Co, Mn and Ti are produced. Cobalt (0.2 to 0.3 %) improves heat resistance and creep properties; nickel improves toughness and titanium affects like grain finer. The main group of this alloy family is the beryllium bronzes with 2 % of Be content due to the highest mechanical properties after the precipitin hardening.

Beryllium bronzes thermal treatment consists of dissolved annealing (700 to 800 °C/1h) and water quenching. The alloy after heat treatment is soft, formable and it can be improved only by artificial aging. Hardening is in progress at temperature from 280 to 300 °C. After the hardening the tensile strength of the alloy is more than 1200 MPa and the hardness 400 HB. By cold forming, applied after the cooling from the annealing temperature, the materials tensile strength can be improved. Beryllium bronzes usage is given by their high tensile strength, hardness, and corrosion resistance which those alloys do not lose, even not in the hardened state. They are used for the good electric conductive springs production; for the equipment which should not sparkling in case of bumping (mining equipment) production; form dies, bearings, etc. (Skočovský et al., 2000, 2006).

2.2.6 Nickel bronzes

Copper and nickel are absolutely soluble in the liquid and in the solid state. Binary alloys are produced with minimal alloying elements content. Complex alloys, ternary or multi components, are suitable for hardening. Nickel bronzes have good strength at normal and also at elevated temperatures; good fatigue limit, they are resistant against corrosion and also against stress corrosion, and they have good wear resistance and large electric resistance.

Binary alloys Cu-Ni with low Ni content (bellow 10 %) are used only limitedly. They are replaced by cheaper Cu alloys. Alloys with middle Ni content (15 to 30 %) have good corrosion resistance and good cold formability. 15 to 20 % Ni containing alloys are used for deep-drawing. Alloys with 25 % Ni are used for coin production and alloys with 30 % Ni are used in the chemical and food industry.

Complex Cu-Ni alloys have a wider usage in the technical praxis compared to the binary alloys. CuNi30Mn with Ni content from 27 to 30 %, Mn content from 2 to 3 % and impurities content bellow 0.6 % is characterized by high strength and corrosion resistance also at elevated temperatures. Because of its electric resistance this alloy is suitable for usage as resistive material till 400 °C. CuNi45Mn constantan is alloy with Ni content form 40 to 46 %, Mn content from 1 to 3 % and impurities content below 0.5 %. From the Cu-Ni alloys, this one has the largest specific electric resistance and it is used for resistive and thermal element material.

Most often the Cu-Ni-Fe-Mn alloys are used. Iron and manganese addition improve the corrosion properties markedly, especially in the seas water and overheated water steam. CuNi30 alloy with iron content in the range from 0.4 to 1.5 % and manganese content from 0.5 to 1.5 % is used for seagoing ships condensers and condensers pipes production. In the new alloys also the niobium as an alloying element is used and the nickel content tends to be decreased because of its deficit. An alloy CuNi10Ge with nickel content from 9 to 10 % and Fe content from 1 to 1.75 % and maximally 0.75 % of Mn, which is used as the material for seagoing ships condensers (Skočovský et al., 2000, 2006).

3. References

Skočovský, P. et al. (2000). *Designing materials* [in Slovak] (1st edition), EDIS, ISBN 80-7100-608-4, Žilina, Slovak republic.

Skočovský, P. et al. (2006). *Material sciences for the fields of mechanical engineering* [in Slovak] (2nd edition), EDIS, ISBN 80-8070-593-3, Žilina, Slovak republic.

Schmidt, R. F. & Schmidt, D. G. (1997) Selection and Application of Copper Alloy Casting, In: *ASM Handbook Volume 2: Properties and Selection: Nonferrous Alloys and Special-Purpose Materials*, pp. 1150-1180, ASM International, ISBN 0-87170-378-5 (v. 2), USA.

Schmidt, R. F., Schmidt, D. G. & Sahoo, M. (1998) Copper and Copper Alloys, In: *ASM Handbook Volume 15: Casting*, pp. 1697-1734, ASM International, ISBN 0-87170-007-7 (v. 1), USA.

http://www.mtfdca.szm.com/subory/med-zliatiny.pdf

http://www.copper.org/education/production.html

http://cst-www.nrl.navy.mil/lattice/struk/a1.html

http://jeanes.webnode.sk/prvky/med/

2

Interaction of Copper Alloys with Hydrogen

I. Peñalva[1], G. Alberro[1], F. Legarda[1],
G. A. Esteban[1] and B. Riccardi[2]
[1]*University of the Basque Country (UPV/EHU),*
Dept. Nuclear Engineering & Fluid Mechanics,
Faculty of Engineering, Bilbao,
[2]*Fusion for Energy, Barcelona,*
Spain

1. Introduction

Copper alloys are well known for their electrical and thermal conductivity, good resistance to corrosion, ease of fabrication and good strength and fatigue resistance. These properties make copper alloys suitable for several electrical and heat-conduction industrial applications. However, many of these industrial processes deal with hydrogen, and the interaction of this gas with copper alloys may affect to their mechanical features. Hydrogen dissolves in all metals to some extent. The dissolved hydrogen in the bulk of the material may change its mechanical properties assisting in its fracture, for example, and leading the material to the so-called hydrogen embrittlement. Therefore, it becomes important to characterise the transport properties of hydrogen in copper alloys as well as their ability to migrate by diffusion through structural walls by interstitial dissolution and trapping. This characterisation allows the improvement of the aforementioned industrial applications.

Lately, copper alloys are being considered as a technical option to construct a pipeline to transport any gaseous fuel including those of high hydrogen content or even pure hydrogen. In relation to this matter, the evaluation of hydrogen migration through the wall of the pipeline and the definition of related fundamental physics are key-issues when performing any risk evaluation because of hydrogen leak capacity. Apart from this question, it is well known the ability of hydrogen to damage copper alloys at high temperatures when they contain oxygen, this problem being directly connected to the ability of hydrogen to migrate through the solid material.

The research in nuclear fusion technology is also highly interested in copper materials. In fact, copper alloys have been selected as structural/heat sink materials that may be used in future fusion reactors like ITER because of their high thermal conductivity, good mechanical properties, thermal stability at high temperature and good resistance to irradiation-induced embrittlement and swelling. In this research area, heat sink/structural materials are subjected to high heat flux and, therefore, must possess a combination of high thermal conductivity and high mechanical strength. Apart from the previous properties, the interaction of hydrogen isotopes with copper alloys that could be part of the in-vessel components of a fusion reactor is of primary importance because it affects to the fuel economy, the plasma stability and the radiological safety of the facility. There are various examples of predictive works trying to establish the time dependant evolution of migration fluxes and fuel inventory within fusion reactor components (Esteban et al., 2004; Meyder et al., 2006) by means of numerical simulation codes that use the hydrogen transport and trapping properties as the main input parameters. Trapping is the process by which dissolved hydrogen atoms remain bound to some specific centres known as "traps" (e.g. inclusions, dislocations, grain boundaries and precipitates). Hence, hydrogen isotopes may be dissolved in trapping or lattice sites of the material. The effect of trapping on hydrogen transport affects to the transport parameters and also to the physical and mechanical properties of the copper alloys involved.

Two of the most promising copper alloys at the present time in several specialised research areas are oxide dispersion strengthened (DS) copper alloys and precipitation hardened (PH) copper alloys (Barabash et al., 2007; Fabritsiev & Pokrovsky, 2005; ITER, 2001; Lorenzetto et al., 2006; Zinkle & Fabritsiev, 1994). This chapter will analyse and compare the experimental hydrogen transport parameters of diffusivity, permeability and Sieverts' constant for the diffusive regime of these kinds of copper alloys. Results can then be extrapolated as general behaviour for similar copper alloys. Trapping properties will also be discussed. Data shown in the chapter will refer to real experimental values for copper alloys obtained by means of the gas evolution permeation technique.

2. Experimental

The gas evolution permeation technique is widely used to characterize the hydrogen transport in metallic materials and, therefore, it turns out to be a suitable technique for the analysis of hydrogen transport in specimens made of different copper alloys. Oxide dispersion strengthened (DS) copper alloys and precipitation hardened (PH) copper alloys have been characterized by means of this experimental method. More precisely, experimental hydrogen transport data are available for a DS copper alloy named GlidCop® Al25 and for a PH-CuCrZr copper alloy named ELBRODUR®.

2.1 Dispersion strengthened and precipitation hardened copper alloys

The GlidCop® Al25 copper alloy is produced by OMG America and contains wt. 0.25 % Al in the form of Al_2O_3 particles. The material is manufactured by means of powder metallurgy using Cu-Al alloy and copper oxide powders. These are mixed and heated to form alumina and then consolidated by hot extrusion. This fabrication method derives in a high density of homogeneously distributed Al_2O_3 nanometric particles within the elongated grain substructure of the material, which are thermally stable and resistant to coarsening so that the grain substructure is resistant to thermal annealing effects (Esteban et al., 2009).

The ELBRODUR® copper alloy is produced by KME-Germany AG. The alloy composition is wt. 0.65 % Cr, wt. 0.05 % Zr and the rest Cu. The fabrication process and the heat treatment consisting of solution annealing (1253 K, 1 h), water quenching and aging (748 K, 2 h) makes possible the presence of nanometric Guinier-Preston zones and incoherent pure Cr particles that provide the material with the high mechanical strength by dislocation motion inhibition. An image of the CuCrZr microstructure is shown in *Figure 1*.

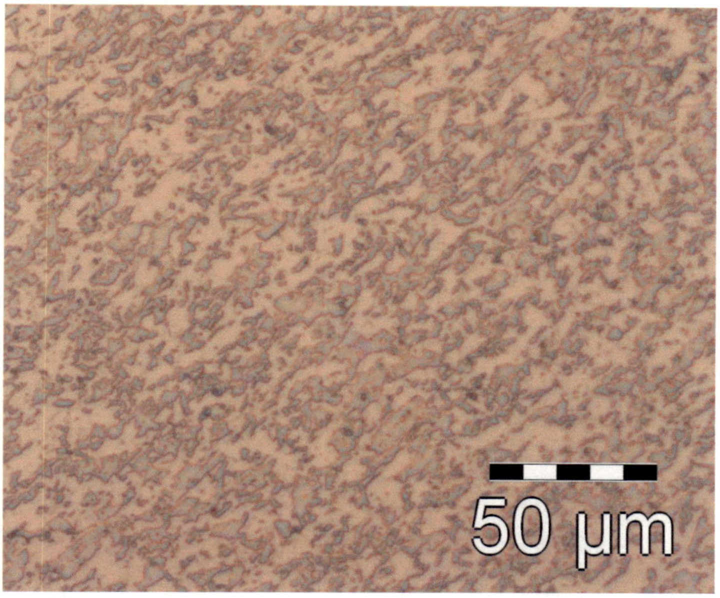

Fig. 1. CuCrZr microstructure.

2.2 Gas evolution permeation technique

A schematic view of a permeation facility is shown in *Figure 2*. The physical principle of the experimental technique entails the gas flux recording that passes through a thin membrane of the material of interest from a high gas pressure region to a low-pressure region at initial vacuum conditions.

The hydrogen migration through the specimen is measured by recording the pressure increase with time in the low-pressure region with two capacitance manometers (Baratron MKS Instr.-USA) P1 and P2 with full scale range of 1000 Pa and 13.33 Pa respectively. An electrical resistance furnace (F) regulated by a PID controller allows to establish the sample temperature within a +/- 1 K precision. The temperature of the specimen is measured by a Ni-Cr/Ni thermocouple inserted into a well drilled in one of the two flanges where the specimen is mounted. The pressure controller (PC) allows the instant exposure of the high-pressure face of the specimen to any desired gas driving pressure, which is measured by means of a high-pressure transducer (HPT).

Before any experimental test is performed with high purity hydrogen (99.9999%), ultra-high vacuum state is reached inside the experimental volumes (up to 10^{-7} Pa) in order to assure the absence of any deleterious species (such as oxygen or water vapour) that may provoke surface oxidation of the specimen (S). There are three ultra-high vacuum pumping units, UHV, composed by a hybrid turbomolecular pump and a primary pump; they pump down the inner volumes of the rig to the desired vacuum level with the help of heating tapes. The vacuum state is checked with three Penning gauges PG in different zones of the facility. A quadrupole mass spectrometer (QMS) is available to check the purity of the gas before and after any experimental test and as an alternative means of testing the quality of the vacuum.

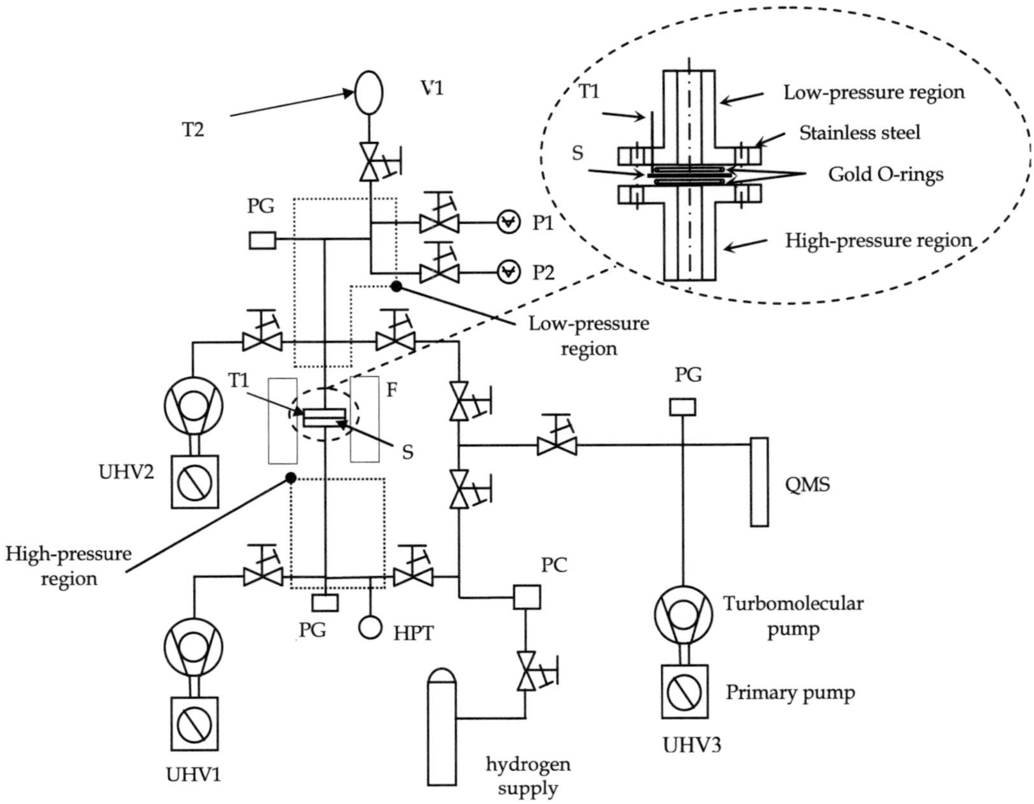

Fig. 2. Schematic view of the permeation facility. PG – Penning gauge; F – furnace; PC – pressure controller; HPT – high-pressure transducer; QMS – quadrupole mass spectrometer; S – specimen; T1, T2 – nickel/chromium-nickel thermocouples; P1, P2 – capacitance manometers; UHV – ultra-high vacuum pumping units, V1 – calibrated volume.

In an individual experimental test, the high driving pressure starts forcing permeation through the high-pressure face of the specimen towards the low-pressure region, where the hydrogen permeation flux rises progressively with time until a steady-state permeation flux is reached. After every experimental permeation run, an expansion of the gas in the low-pressure region is performed to a calibrated volume (V1) in order to convert pressure values into permeated gas amount, or alternatively, the speed of pressure increase into permeation flux. The modelling of the pressure increase $p(t)$ due to the gas permeation towards the low-pressure region (a typical experimental permeation curve is shown in *Figure 3*) makes possible to obtain the hydrogen transport properties of the copper alloy: permeability (Φ), diffusivity (D) and Sieverts´ constant (K_S).

The permeated flux under diffusive regime for every temperature depends on the thickness of the sample, the values of the loading pressure and the permeability of the gas (Φ). This transport parameter defines the gas-material interaction. Diffusion is a physical property that allows the flux of a gas through the bulk of a solid material due to, in this case, a concentration gradient of the dissolved hydrogen. The gas flux in the bulk of the material depends on the concentration gradient and on the temperature. The proportionality between the flux and the concentration gradient is called diffusivity (D) and is directly related to the kinetics of the system in order to reach the equilibrium by means of diffusion. Finally, Sieverts´ constant (K_S) is directly related to the solubility of the gas in the solid and can be derived from the values of diffusivity and permeability.

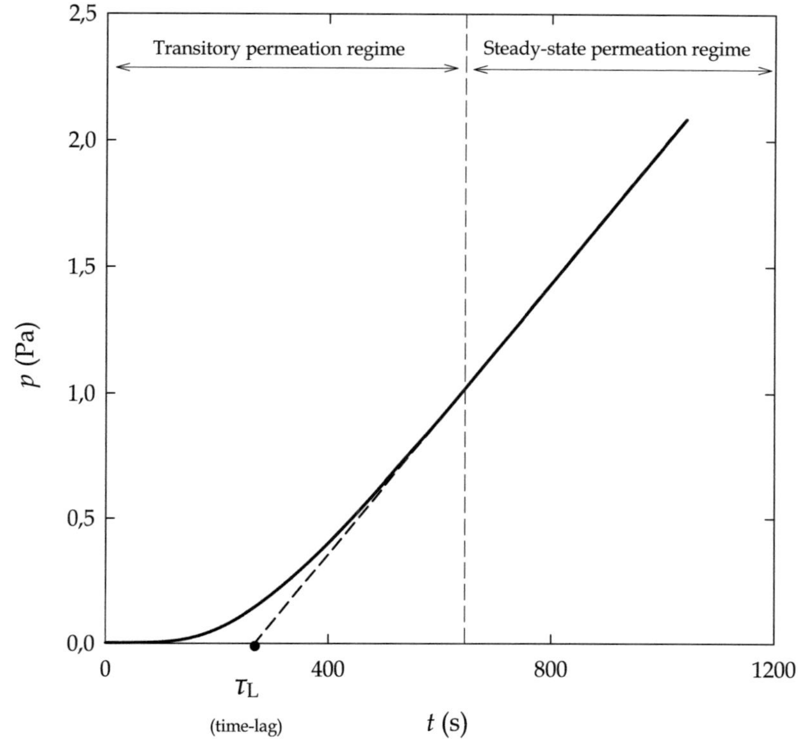

Fig. 3. Experimental permeation curve: transitory permeation regime and steady-state permeation regime. Definition of the time-lag.

3. Theory

Typical bulk parameters for the study of hydrogen transport in metal lattices are the diffusivity (D), the Sieverts´ constant, (K_S), and the permeability (Φ) (Alberici & Tominetti, 1995). The diffusivity is related to the diffusing flux in a metallic matrix, J, and the gradient of the gas concentration in the matrix, ∇c, by the first Fick's law:

$$J = -D \cdot \nabla c \tag{1}$$

where is easy to see that D is linked to the migration velocity of the gas in the material.

In this equation, taking into account a homogeneous bulk, D will be supposed to be uniform and constant throughout the material volume and it only depends on the absolute temperature, T, by an Arrhenius relationship:

$$D = D_0 \cdot \exp(-E_d / R \cdot T) \quad (2)$$

where E_d is the diffusion activation energy, which is always positive.

Experimentally it is found that hydrogen dissolves atomically in metal lattices; the proportionality between the atomic gas concentration in the bulk volume, c, and the square root of the equilibrium gas pressure outside the bulk, $p^{1/2}$, is known as the Sieverts´ constant:

$$K_S = c / \sqrt{p} \quad (3)$$

It is interesting to note that Sieverts´ constant also shows an Arrhenius dependence on the temperature:

$$K_S = K_{S,0} \cdot \exp(-E_s / R \cdot T) \quad (4)$$

where E_S is the activation energy for solution, which can be either positive or negative.

Permeability (Φ) is given by means of Richardson's law that states a linear relation between D and K_S:

$$\phi = K_S \cdot D \quad (5)$$

From Eq. 5, it is obvious that permeability also follows an Arrhenius behaviour like D and K_S, with an activation energy of permeation which is the sum of E_d and E_s:

$$\phi = \phi_0 \cdot \exp(-(E_d + E_s) / R \cdot T) \quad (6)$$

All the processes involved in the interaction between hydrogen and the metallic material, either on the surface or in the bulk, may be explained by the analysis of the different potential energy levels acquired by the hydrogen atom/molecule in the immediacy of, and within the metal (Esteban et al., 1999). These energy levels are summarised in *Figure 4* (Möller, 1984). Out of the material, hydrogen is in the molecular form: the solid line refers to atomic hydrogen and the broken line to molecular hydrogen. All the energy increments and decrements depicted in *Figure 4* define the hydrogen behaviour within, and in the vicinity of the solid metal and explain observed physical processes. The dissociation energy, E_{di}, is the amount of energy needed for splitting a hydrogen molecule into two atoms. The chemisorption energy, E_{ch}, refers to the chemical binding established between atomic hydrogen and metallic atoms. The adsorption energy, E_{ad}, is the energy barrier hydrogen has to surmount in order to access to a chemisorption site and it depends on the surface condition. The solution energy, E_s, is the energy difference between a free atom and a dissolved one and depending on the sign of this energy the material is characterised as endothermic, $E_s > 0$, or exothermic, $E_s < 0$. The diffusion energy, E_d, is the barrier the diffusing atom has to surmount in order to pass, within the lattice, from one solution site to another. The trapping energy, E_t, is the potential well to which a hydrogen atom remains bound when interacting with the potential trapping sites. ΔE states for the energy difference between a normal solution site and a trapping site, (E_s – E_t). Finally, E_c states for the energy difference when comparing potential barriers between normal solution sites and a trapping site.

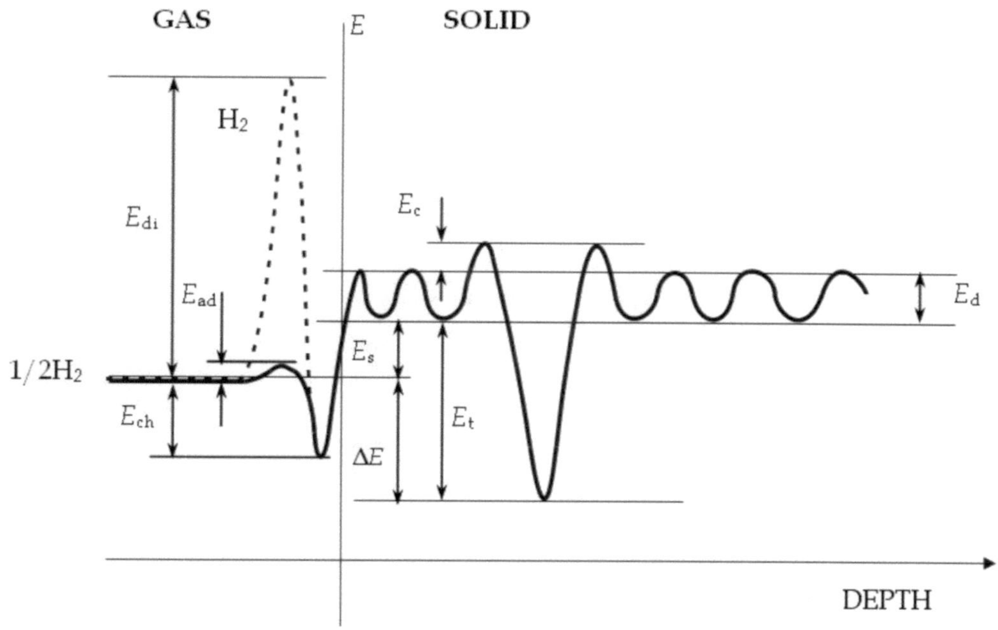

Fig. 4. Potential energy distribution in a metal (Möller, 1984).

Hydrogen isotope transport through material may be limited either by gas interstitial diffusion through the bulk (diffusion-limited regime) or by the physical-chemical reactions of adsorptive dissociation and desorptive recombination occurring on the surface of the solid material (surface-limited regime). The objective of this experimental task is usually to characterize the diffusion-limited regime instead of the surface-limited regime, the second one being only relevant when any kind of impurities or oxides are present on the surface of the material.

"Trapping" is the process by which dissolved hydrogen atoms remain bound to some specific centres known as "traps" (e.g. inclusions, dislocations, grain boundaries and precipitates). Hence, hydrogen isotopes may be dissolved in trapping or lattice sites of the material. The effect of trapping on hydrogen transport is, on the one hand, the increase in the gas absorbed inventory, i.e. the increase in the effective Sieverts' constant ($K_{S,eff}$) with respect to the aforementioned lattice Sieverts' constant (K_S). On the other hand, the dynamics of transport becomes slower, i.e. the decrease of the effective diffusivity (D_{eff}) with respect to the aforementioned lattice diffusivity (D). As a result, the Arrhenius temperature dependence of the parameters remains modified as follows, according to Eqs. (2) and (4) for diluted solutions (Oriani R.A., 1970):

$$D_{eff} = \frac{D}{1 + \frac{N_t}{N_l}\exp(E_t / R \cdot T)} \tag{7}$$

$$K_{S,eff} = K_S \cdot \left(1 + \frac{N_t}{N_l}\exp(E_t / R \cdot T)\right) \tag{8}$$

D_0 and $K_{S,0}$ being the pre-exponential lattice diffusivity and pre-exponential lattice Sieverts' constant, and E_d, E_s the diffusion and solution energies, respectively. N_t (m^{-3}) is the trap sites concentration, N_l (m^{-3}) is the lattice dissolution sites concentration and E_t the trapping energy.

When the individual effective parameters for each experimental temperature have been obtained, another fitting routine is separately run with Eqs. (2), (4), (7) and (8) for the lattice parameters D_0, E_d, $K_{S,0}$ and E_s and trapping parameters E_t and N_t over the correspondent temperature range of influence. The value of $8.5 \cdot 10^{28}$ m^{-3} is taken for the density of solution sites into the lattice N_l, assuming that the copper alloy is close to a fcc structure where hydrogen occupies only the octahedral interstitial positions (Vykhodets et al., 1972).

The effective transport parameters of diffusivity (D_{eff}) and permeability (Φ) are evaluated for each temperature by modelling the experimental permeation curves obtained for every individual test. The Sieverts' constant ($K_{S,eff}$) is derived from the definition of permeability that states the relationship amongst the three transport parameters:

$$\Phi = D_{eff} \cdot K_{S,eff} \tag{9}$$

A subsequent analysis of the Arrhenius dependence of these transport parameters with temperature enables the obtaining of the characteristic transport parameters of trapping energy (E_t) and density of traps (N_t).

The obtaining of the theoretical expression for the pressure increase with time in the low-pressure region as a function of the previous transport parameters is briefly explained hereafter.

The specimens are thin discs with a very high ratio of the circular surface exposed to the gas in relation to the length of the diffusion path through the bulk of material. This is the reason why the problem can be modelled by an infinite slab with gas diffusion occurring in the direction perpendicular to the surface of the specimen.

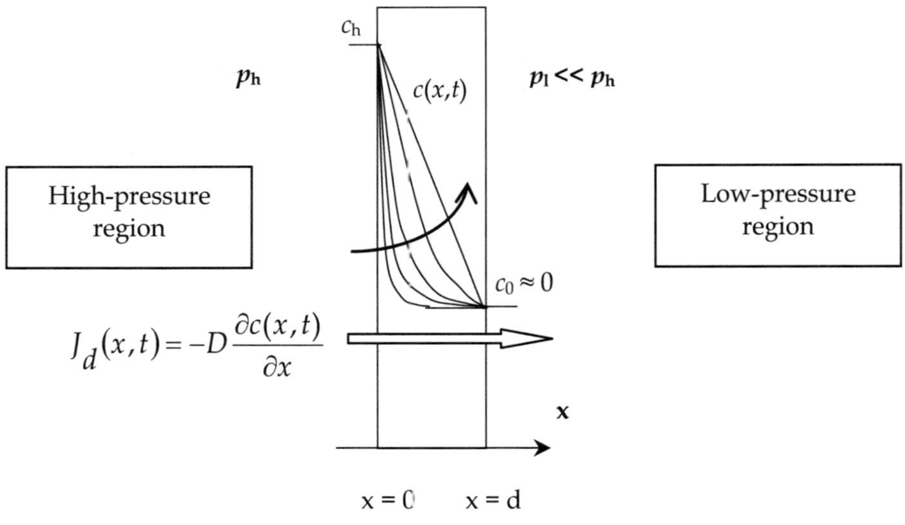

Fig. 5. Scheme of the permeation process through a 1-D slab. p_h – high-pressure; p_l – low-pressure; d – thickness of the slab; $J_d\,(x,t)$ – diffusive flux; $c(x,t)$ – gas concentration.

A scheme of the gas transport through a sheet of material with a certain thickness (d) is shown in *Figure 5*. That specimen is exposed on one side to a certain gas driving pressure (p_h), whereas the other side is left under vacuum conditions (i.e. very low pressure p_l).

The hydrogen concentration ($c(x,t)$) at each position (x coordinate) and each time (t) may be determined by solving the second Fick's law in the one dimension slab:

$$\frac{\partial c(x,t)}{\partial t} = D_{eff} \frac{\partial^2 c(x,t)}{\partial x^2} \tag{10}$$

The boundary conditions being the following:

- 1st condition: $c(x=0,t) = c_h$, from the beginning, in the region closest to the surface, the gas concentration acquires the final equilibrium value in the saturation state given by Sieverts' law,

$$c_h = K_{S,eff} \cdot p_h^{0.5} \tag{11}$$

- 2nd condition: $c(x=d,t) = 0$, the gas concentration in the low-pressure side is negligible in comparison to c_h; i.e. p_l negligible in comparison to p_h,

$$c_0 = K_{S,eff} \cdot p_l^{0.5} \tag{12}$$

The initial condition is $c(x>0,t=0)=0$; at the beginning of the test, the specimen is under vacuum conditions without any amount of hydrogen dissolved into the material.

The analytical solution of the Eq. (10) with the previous boundary and initial conditions is (Carslaw & Jaeger, 1959):

$$c(x,t) = c_h\left(1-\frac{x}{d}\right) - \frac{2c_h}{\pi}\sum_{n=1}^{\infty}\frac{1}{n}\sin\left(\frac{n\cdot\pi\cdot x}{d}\right)\exp\left(-D_{eff}\frac{n^2\cdot\pi^2}{d^2}t\right) \tag{13}$$

The resultant flux to the low-pressure region can be evaluated as:

$$J(x=d,t) = -D\left.\frac{\partial c(x,t)}{\partial t}\right|_{x=d} = \frac{D_{eff}\cdot K_{S,eff}\cdot p_h^{0.5}}{d}\left[1+2\sum_{n=1}^{\infty}(-1)^n\exp\left(-D\frac{n^2\cdot\pi^2}{d^2}t\right)\right] \tag{14}$$

The total gas inventory ($I(t)$) permeated to the low-pressure region is evaluated by accounting for all the gas flux released during the considered time period (t) and taking into account the surface area of the specimen (A_s):

$$I(t) = A_s\int_0^t J(d,t')\,dt' = \frac{\Phi\cdot p_h^{0.5}}{d}A_s\cdot t - \frac{\Phi\cdot p_h^{0.5}\cdot d}{6\cdot D_{eff}}A_s - \frac{2\cdot\Phi\cdot p_h^{0.5}\cdot d}{6\cdot D_{eff}}A_s\sum_{n=1}^{\infty}\frac{(-1)^n}{n^2}\exp\left(-D_{eff}\frac{n^2\cdot\pi^2}{d^2}t\right) \tag{15}$$

Taking into account the ideal gas approximation, pressure increment with time in the low-pressure region due to this amount of gas is:

$$p(t)=\frac{\mathrm{R}\cdot T_{eff}}{V_{eff}}\left[\frac{\Phi\cdot p_{\mathrm{h}}^{0.5}}{d}A_{\mathrm{s}}\cdot t-\frac{\Phi\cdot p_{\mathrm{h}}^{0.5}\cdot d}{6\cdot D_{\mathrm{eff}}}A_{\mathrm{s}}-\frac{2\cdot\Phi\cdot p_{\mathrm{h}}^{0.5}\cdot d}{6\cdot D_{\mathrm{eff}}}A_{s}\sum_{n=1}^{\infty}\frac{(-1)^{n}}{n^{2}}\exp\left(-D_{\mathrm{eff}}\frac{n^{2}\cdot\pi^{2}}{d^{2}}t\right)\right] \quad (16)$$

Where the V_{eff} is the effective volume where the permeated gas is retained, T_{eff} is the temperature of the volume and R is the ideal gas constant (8.314 J · K^{-1} · mol^{-1}). The volume V_{eff} is precisely measured in each experimental permeation test by performing gas expansion to a calibrated volume.

When imposing a very large period of time ($t\rightarrow\infty$) in the previous expression the evolution of pressure with time for the steady-state permeation regime is obtained:

$$p_{\infty}(t)=\frac{\mathrm{R}\cdot T_{\mathrm{eff}}}{V_{\mathrm{eff}}}\left(\frac{\Phi\cdot p_{\mathrm{h}}^{0.5}}{d}A_{s}\cdot t-\frac{\Phi\cdot p_{\mathrm{h}}^{0.5}\cdot d}{6\cdot D_{\mathrm{eff}}}A_{s}\right) \quad (17)$$

This expression corresponds to the steady-state flux,

$$J_{\infty}=\frac{\Phi\cdot p_{\mathrm{h}}^{0.5}}{d} \quad (18)$$

obtained from Eq. (17); this is the linear tendency shown in *Figure 3* on the right-hand side. When the straight line is extended down to cross the abscise axis in the time co-ordinate a characteristic time known as time-lag is obtained:

$$\tau_{\mathrm{L}}=\frac{d^{2}}{6\cdot D_{\mathrm{eff}}} \quad (19)$$

The value of permeability (Φ) can be derived from the slope of the straight line in steady-state permeation regime (Eq. (17)) and the effective diffusivity (D_{eff}) can be derived from the value of the time-lag. Nevertheless, a non-linear least-squares fitting to all the experimental points of each single test has been preferred with the general expression (Eq. (16)) in both the steady-state region and the transitory region by considering the permeability (Φ) and the diffusivity (D_{eff}) as the fitting parameters.

In any individual permeation test, the gas is on contact with a solid surface and the hydrogen concentration profile through the sample thickness rises, becoming linear and stable after certain period of time. In that final permeation process the relationship between steady-state flux (J_{∞}) and the loading pressure (p_{h}) will be different depending whether the transport regime is diffusion-limited or surface-limited (Esteban et al., 2002):

$$J_{\infty}=\frac{\Phi}{d}\cdot p_{\mathrm{h}}^{0.5}\ \text{(diffusion-limited)} \quad (20)$$

$$J_{\infty}=\frac{1}{2}\cdot\sigma\, k_{1}\cdot p_{\mathrm{h}}\ \text{(surface-limited)} \quad (21)$$

where σk_1 is the adsorption rate constant. The experimental confirmation of one of these relationships is a method to decide the type of transport regime for modelling the experimental tests.

4. Results and discussion

This section reviews the available data in literature for oxide dispersion strengthened (DS) copper alloys and precipitation hardened (PH) copper alloys. Results regarding interaction of these alloys with hydrogen are compared in relation to base material, Cu. Punctual experimental values are shown only for the ELBRODUR® alloy (not published), whereas Arrhenius regressions of the transport parameters are compiled for all the alloys.

Individual permeation tests have been carried out for the aforementioned copper alloys, GlidCop® Al25 (Esteban et al., 2009) and ELBRODUR®, with temperatures ranging from 573 K to 793 K and using loading pressures ranging from 10^3 Pa to $1.0 \cdot 10^5$ Pa. Additionally, data for base material, Cu, (Reiter et al., 1993) and for a similar PH-CuCrZr alloy (Serra & Perujo, 1998) named ELBRODUR-II hereafter to distinguish from the material analysed, are also available. These results for the hydrogen transport parameters in copper alloys are summarised and discussed in the next paragraphs.

In relation to the permeation tests carried out for GlidCop® Al25 and ELBRODUR® copper alloys, the evaluation of the diffusive transport parameters has been assured because no surface effect has become relevant within the whole group of individual tests. This fact has been proved by studying the evolution of the experimental steady-state flux (J_∞) with driving pressure (p_h) at the same temperature.

In the case of the ELBRODUR® copper alloy, a set of 9 permeation tests has been performed at the same temperature (688 K) with different loading pressures (p_h) in order to study the type of hydrogen transport regime. These results are shown in *Figure 6*. The exponential relationship between the steady-state hydrogen flux (J_∞) and the loading pressure (p_h) has a power of n = 0.52, which is close to 0.5 (pure diffusion-limited regime) and far from 1.0 (pure surface-limited regime) (Eqs. (20) and (21), respectively).

Individual transport parameters of effective diffusivity (D_{eff}), permeability (Φ) and effective Sieverts' constant ($K_{S,eff}$) have been obtained at different temperatures by modelling the corresponding individual permeation tests, both for GlidCop® Al25 (Esteban et al., 2009) and ELBRODUR®.

The dependence of the transport parameters on temperature for the ELBRODUR® copper alloy is shown in *Figure 7* (permeability), *Figure 8* (diffusivity) and *Figure 9* (Sieverts´ constant), together with the results obtained for the aforementioned reference copper alloys (Esteban et al., 2009; Reiter et al., 1993; Serra & Perujo, 1998). The Arrhenius parameters are obtained by fitting the individual experimental values to the tendencies given by Eqs. (2), (4), (7) and (8), resulting:

$$\Phi\ (\mathrm{mol \cdot m^{-1} \cdot Pa^{-0.5} \cdot s^{-1}}) = 2.38 \cdot 10^{-7} \cdot \exp(-73.9\ (\mathrm{kJ \cdot mol^{-1}}) / \mathrm{R} \cdot T)$$

$$D\ (\mathrm{m^2 \cdot s^{-1}}) = 3.55 \cdot 10^{-5} \cdot \exp(-65.5\ (\mathrm{kJ \cdot mol^{-1}}) / \mathrm{R} \cdot T)$$

$$K_S\ (\mathrm{mol\cdot m^{-3}\cdot Pa^{-0.5}}) = 6.71\cdot 10^{-3}\cdot \exp(-8.4\ (\mathrm{kJ\cdot mol^{-1}})/\mathrm{R}\cdot T)$$

The trapping parameters are $N_t = 3.7 \cdot 10^{24}$ m^{-3} and $E_t = 51.2$ kJ mol^{-1}.

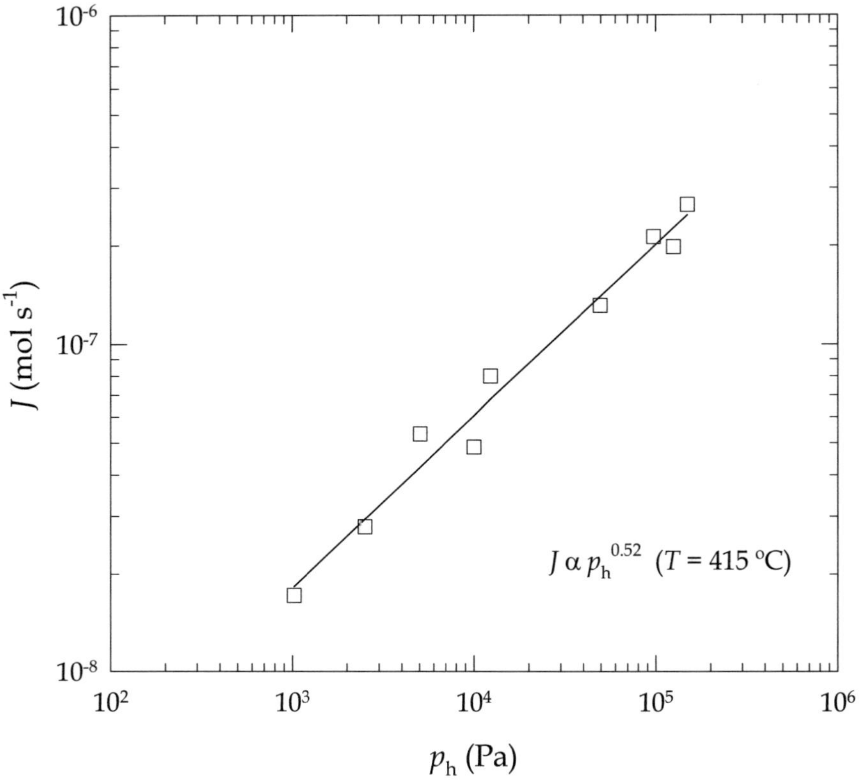

Fig. 6. Experimental hydrogen permeation steady-state flux in PH-CuCrZr alloy (ELBRODUR®) at 688 K and different driving pressures ranging from 10^3 to $1.5 \cdot 10^5$ Pa.

The Arrhenius pre-exponentials and the activation energies of hydrogen transport parameters together with the trapping parameters that have been plotted in *Figs. 7-9* are shown in *Table 1*.

Material (curve)	Φ_0	E_Φ	D_0	E_d	K_{S0}	E_S	N_t	E_t	T
ELBRODUR (1)	$2.38\cdot10^{-7}$	73.9	$3.55\cdot10^{-5}$	65.5	$6.71\cdot10^{-3}$	8.4	$3.7\cdot10^{24}$	51.2	593-773
GlidCop® Al-25 (2) (Esteban et al., 2009)	$5.87\cdot10^{-7}$	80.6	$5.70\cdot10^{-5}$	76.8	0.006	3.7	$3.1\cdot10^{22}$	75.4	573-793
ELBRODUR-II (3) (Serra & Perujo, 1998)	$5.13\cdot10^{-7}$	79.8	$5.70\cdot10^{-7}$	41.2	0.90	38.6	-	-	553-773
Cu (4) (Reiter et al, 1993)	$6.60\cdot10^{-6}$	92.6	$6.60\cdot10^{-7}$	37.4	5.19	55.2	-	-	470-1200

Table 1. Experimental hydrogen transport parameters for reference copper alloys; Φ_0 in mol·m^{-1}·Pa$^{-0.5}$·s^{-1}, E_Φ, E_d, E_s and E_t in kJ· mol^{-1}, D_0 in m^2·s^{-1}, K_{S0} in mol· m^{-3}·Pa$^{-0,5}$, N_t in m^{-3} and T in K.

There exists a marked difference between the transport parameters obtained in the PH-CuCrZr alloy (ELBRODUR®) in relation to the corresponding ones for the base material (Cu) (Reiter et al, 1993), DS copper alloy GlidCop® Al25 (Esteban et al., 2009), and a similar PH-CuCrZr alloy (Serra & Perujo, 1998) named ELBRODUR-II. There is a different metallurgical composition in the Zr content and a slight difference in the thermal treatment of both the ELBRODUR alloys. Moreover, in the work performed with ELBRODUR-II (Serra & Perujo, 1998) the effect of hydrogen trapping was not envisaged.

In the case of the transport property of permeability (*Figure 7*) the result obtained for the PHCuCrZr alloy ELBRODUR® is congruent with the results obtained in other reference Cu alloys. The permeation energy, 73.9 kJ/mol, preserves a similar value to those of the other Cu alloys (80.6 kJ/mol in GlidCop® Al25 and 79.8 kJ/mol in ELBRODUR-II) and it is slightly lower than that of the pure Cu (92.6 kJ/mol).

The aforementioned similar results in the four different materials are reasonable because permeability is a property describing the steady-state hydrogen migration through lattice with no influence of the trapping effect and the particular microstructural defects of each material; i.e. when enough period of time passes, hydrogen concentration dependence on depth adopts the final linear profile (see *Figure 5*) when trapping and detrapping (the inverse process) have reached equal equilibrium rates that cancel each other.

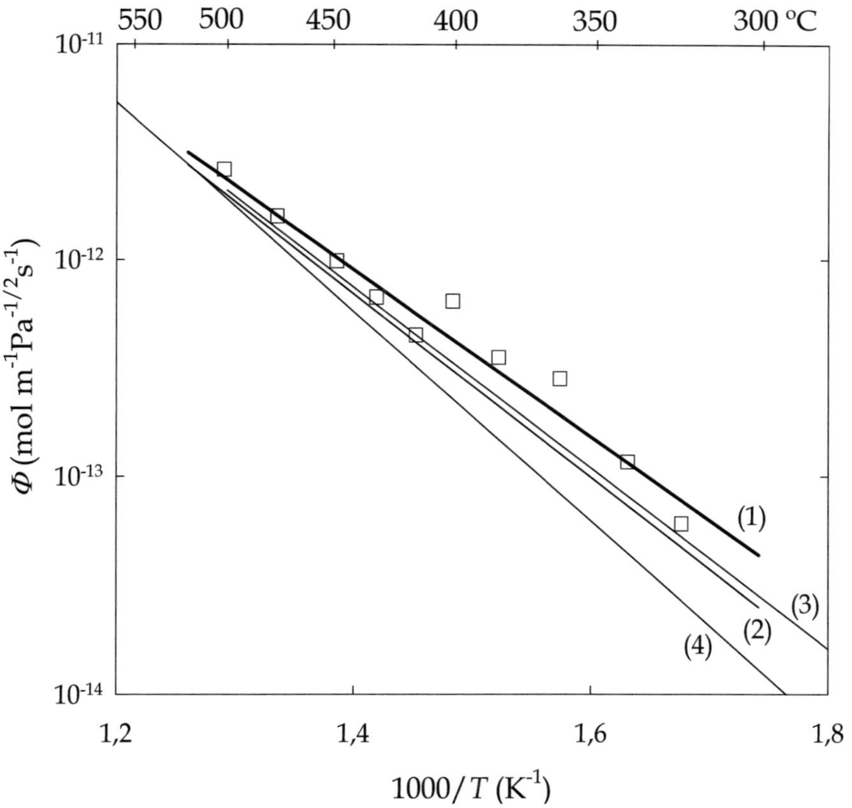

Fig. 7. Hydrogen permeability in PH-CuCrZr ELBRODUR® alloy compared with reference copper alloys: (1) ELBRODUR®, (2) GlidCop® Al25, (3) ELBRODUR-II, (4) pure Cu.

The influence of microstructural defects of the material acting as strong trapping sites for hydrogen absorption, can be observed in transport properties such as diffusivity (D_{eff}) and Sieverts' constant ($K_{S,eff}$). The dependence of the hydrogen diffusivity in PH-CuCrZr ELBRODUR® alloy and DS-GlidCop® Al25 alloy with temperature (shown in *Figure 8*) evidences the influence of trapping that provokes a general decrease in the diffusivity; i.e. the kinetics of migration becomes slower because trapping and detrapping processes impede the free flow of interstitial atoms through lattice solution sites. This effect becomes more pronounced as the temperature is lower (the vibration state of the hydrogen atom is weaker and the high trapping energy well is more effective for hydrogen trapping). At high temperatures, the diffusivity tends to approximate asymptotically to the behaviour of the base material Cu (curve 4) when the trapping effect becomes negligible. The alloys exhibit high values of diffusion energy (65.5 kJ/mol and 76,8 kJ/mol) and a marked influence of the trapping phenomenon with high values of trapping energies (51.2 kJ/mol and 75.4 kJ/mol). In the case of the in PH-CuCrZr ELBRODUR® alloy, the abundant hydrogen trapping sites in this material may be identified with the nanometric Guinier-Preston zones, incoherent pure Cr particles or extensive precipitates like Cu_4Zr characteristic of this kind of alloy (Edwards et al., 2007). This behaviour is analogous to the trapping phenomena described in Gildcop Al25 (Esteban et al, 2009), where the presence of nanometric Al_2O_3 provoked a massive hydrogen trapping phenomenon even more effective than in the PH-CuCrZr alloy.

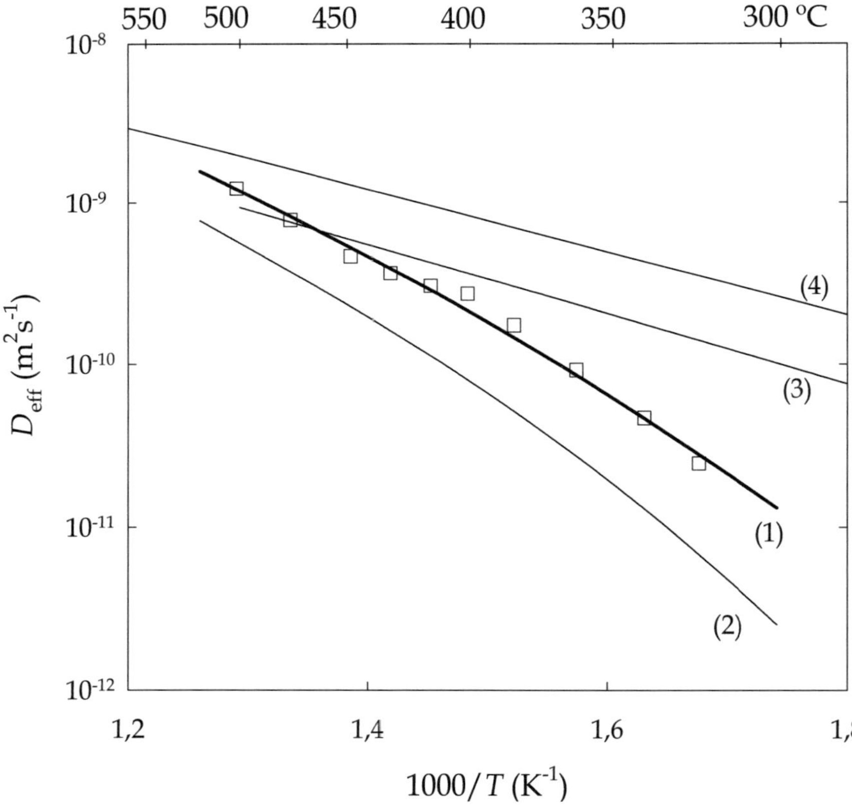

Fig. 8. Hydrogen diffusivity in PH-CuCrZr ELBRODUR® alloy compared with reference copper alloys: (1) ELBRODUR®, (2) GlidCop® Al25, (3) ELBRODUR-II, (4) pure Cu.

The hydrogen Sieverts' constants for the PH-CuCrZr ELBRODUR® alloy and the DS-GlidCop® Al25 alloy are shown in *Figure 9* in comparison to the base material, Cu. All over again, a marked trapping effect in hydrogen Sieverts' constant (i.e. solubility) has been observed throughout the whole temperature range for both alloys (curves 1 and 2). At low temperatures, hydrogen remains trapped into the defects of material exceeding the prediction made by the consideration of normal interstitial lattice sites of the base material Cu (curve 4). The interstitial lattice dissolution remains endothermic but with a low value of the dissolution energy for both alloys (8.4 kJ/mol and 3.7 kJ/mol). The trapped hydrogen specie becomes so important at low temperature that the effective Sieverts' constant behaves as an effective endothermic tendency.

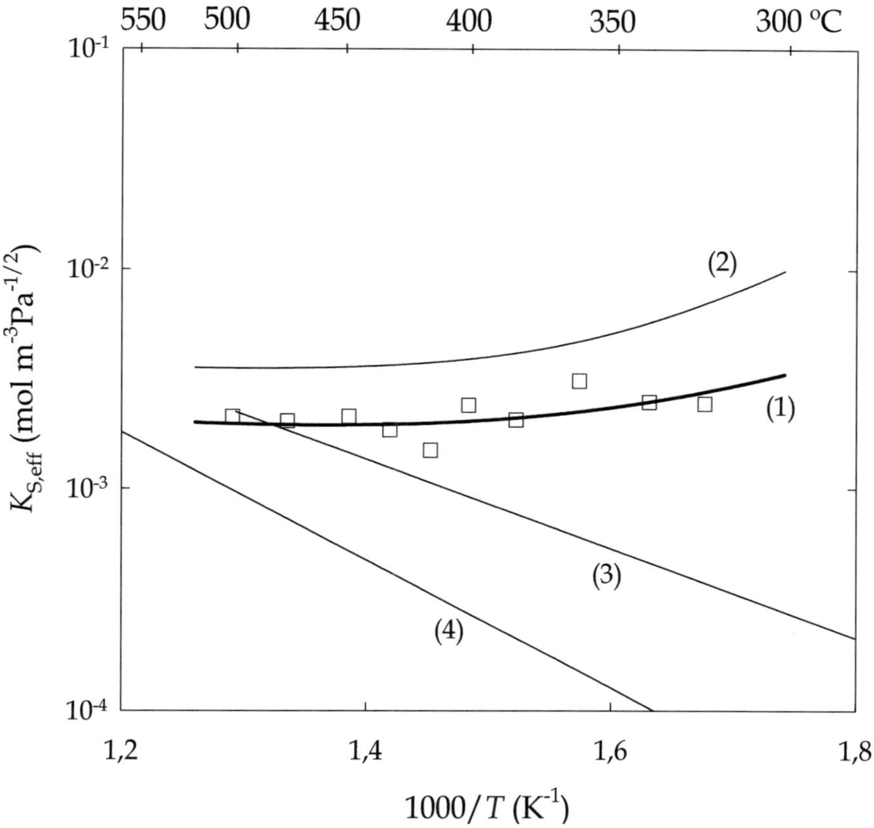

Fig. 9. Hydrogen Sieverts´ constant in PH-CuCrZr ELBRODUR® alloy compared with reference copper alloys: (1) ELBRODUR®, (2) GlidCop® Al25, (3) ELBRODUR-II, (4) pure Cu.

The explanation of these particular tendencies may be found in the presence of high density nanosized defects in the materials. In the case of PH-CuCrZr ELBRODUR® alloy, the hydrogen interstitial atoms may remain trapped in the interface of the Guinier-Preston zones, incoherent Cr particles or precipitates, increasing the solubility and slowing down the transport through the lattice of the material (i.e. a lower effective diffusivity). In the case of the DS-GlidCop® Al25 alloy, the same effect can be attributed to the hydrogen inventory trapped in the nanosized Al_2O_3 particles. Furthermore, this phenomenon has been experimentally identified in other kind of materials like oxide dispersion strengthened (ODS) reduced activation ferritic martensitic (RAFM) steels where nanoparticles of yttria Y_2O_3 provoked an analogous effect (Esteban et al., 2007).

The effect of nanosized inclusions has an obvious successful effect in the improvement of thermal-mechanical properties of copper alloys. However, the effect of the increase of hydrogen isotope inventory retention needs to be taken into account. This effect can be extremely important in particular cases. In fusion reactor materials, for example, it should be taken into account when choosing the structural and heat-sink materials of the fusion reactor where the hydrogen isotope inventory has to be controlled with special attention when considering fuel balance economy or radiological safety issues. When choosing materials for pipelines that will transport gaseous fuels including those with high hydrogen content or even pure hydrogen, the observed hydrogen trapping should be taken into account as long as it may degrade its mechanical properties. On the other hand, electrical characteristics may also be affected by hydrogen trapping phenomenon (Lee K. & Lee Y.K., 2000).

5. Conclusion

The gas permeation technique has been used in order to characterise two copper alloys proposed for high heat flux components: an oxide dispersion strengthened (DS) copper alloy named GlidCop® Al25, and a precipitation hardened (PH) copper alloy named ELBRODUR®. The hydrogen diffusive transport parameters have been obtained and discussed in relation to the particular microstructure of each copper alloy. The hydrogen trapping phenomenon has resulted to be present throughout the whole experimental temperature provoking an increase of hydrogen Sieverts' constant and decrease of diffusivity. The permeability values remained close to the values of the base material, i.e. pure Cu, and the other reference copper alloys. The analogy of the experimental results obtained with other materials with nanosized inclusions, confirms the high ability of these kinds of material to trap hydrogen isotopes at low temperatures. This should be monitored with special care for applications where hydrogen trapping may modify the physical properties of copper alloys.

6. Acknowledgment

This work has been funded by the Spanish Ministry of Science and Education (Ref. ENE2005-03811) with an ERDF proportion. The authors would also like to thank the FEMaS Coordinated Action project for the support in knowledge exchange among different research groups.

7. References

Barabash V., (the ITER International Team), Peacock, A., Fabritsiev, S., Kalinin, G., Zinkle S., Rowcliffe, A., Rensman, J.-W., Tavassoli, A.A., Marmy, P., Karditsas P.J., Gillemot, F. & Akiba, M. (2007). Materials Challenges for ITER – Current Status and Future Activities. *Journal of Nuclear Materials*, Vol. 367-370, Part 1 (August 2007), pp. 31-32, ISSN-0022-3115

Carslaw, H. S. & Jaeger, J. C. (1959) *Conduction of Heat in Solids*, (2nd Edition), Clarendon Press, ISBN-0-19-853368-3, Oxford.

Edwards, D. J., Singh B. N., & Tähtinen S. (2007). Effect of heat treatments on precipitate microstructure and mechanical properties of a CuCrZr alloy. *Journal of Nuclear Materials*, Vol. 367-370, Part 2 (August 2007), pp. 904-909, ISSN-0022-3115

Esteban, G. A., Sedano, L. A., Perujo A., Douglas K., MAncinelli B., Ceroni P.l., Cueroni G.B. (1999). Hydrogen Transport Parameters and Trapping Effects in the Martensitic Steel Optifer-IVb. Report EUR 18995 EN (1999).

Esteban, G. A., Perujo A, Sedano, L. A., Legarda, F. Mancinelli, B. & Douglas, K. (2002). Diffusive transport parameters and surface rate constants of deuterium in Incoloy 800. *Journal of Nuclear Materials*, Vol. 300, Iss. 1 (January 2002), pp. 1-6, ISSN-0022-3115

Esteban, G. A., Perujo A, & Legarda, F. (2004). Tritium Management in the First-Wall Materials of A-DC and TAURO Blankets. *Journal of Nuclear Materials*, Vol. 335, Iss. 3 (December 2004), pp. 353-358, ISSN-0022-3115

Esteban, G. A., Peña, A., Legarda, F. & Lindau, R. (2007). Hydrogen Transport and Trapping in ODS-EUROFER. *Fusion Engineering and Design*, Vol. 82, Iss. 15-24 (October 2007), pp. 2634-2640, ISSN-0920-3796

Esteban, G. A., Alberro, G., Peñalva, I., Peña, A., Legarda, F. & Riccardi, B. (2009). Hydrogen Transport and Trapping in the GlidCop® Al25 IG Alloy. *Fusion Engineering and Design*, Vol. 84, Iss. 2-6 (June 2009), pp. 757-761, ISSN-0920-3796

Fabritsiev, S.A. & Pokrovsky, A.S. (2005). Effect of high doses of neutron irradiation on physico-mechanical properties of copper alloys for ITER applications. *Fusion Engineering and Design*, Vol. 73, Iss. 1 (April 2005), pp. 19-34, ISSN-0920-3796

ITER Doc. (2001). *ITER Materials Assessment Report (MAR)*, ITER Doc. G 74 MA 10 01-07-11 W0.2 (internal project document distributed to the ITER Participants).

Lee K. & Lee Y.K., (2000), Irreversible hydrogen effects on resistivity of sputtered copper films. *Journal of Materials Science*, Vol. 35. (May 2000), pp.6035-6040

Lorenzetto, P., Peacock, A., Bobin-Vastra, I., Briottet L., Bucci, P., Dell'Orco, G., Ioki, K., Roedig, M. & Sherlock, P. (2006). EU R&D on the ITER First Wall. *Fusion Engineering and Design*, Vol. 81, Iss. 1-7 (February 2006), pp. 355-360, ISSN-0920-3796

Meyder, R., Boccaccini, L. V. & Bekris, N. (2006) Tritium analysis for the European HCPB TBM in ITER, Proceedings of IEEE/NPSS 21st Symposium on Fusion Engineering, ISBN- 0-4244-0150-X, pp. 267-270, Knoxville, TN USA, September 2005

Möller W., (1984). Physics of Plasma-Wall Interaction in Controlled Fusion, NATO AISI series, p. 439, (1994)

Oriani R.A. (1970). The Diffusion and Trapping of Hydrogen in Steel, *Acta Metallurgica*, Vol. 18, (January 1970) , pp. 147-157

Reiter, F., Forcey, K.S. & Gervasini, G. (1993). A Compilation of Tritium-Material Interaction Parameters in Fusion Reactor Materials. Report EUR 15217 EN (1993).

Serra, E. & Perujo, A. (1998). Hydrogen and Deuterium Transport and Inventory Parameters in a Cu-0.65Cr-0.08Zr Alloy for Fusion Reactor Applications, *Journal of Nuclear Materials*, Vol. 258-263, Part 1 (October 1998), pp. 1028-1032, ISSN-0022-3115

Vykhodets, V. B., Geld, P. V., Demin, V. B. Men, A. N., Murtazin, I. A. & Fishman A. Ya. (1972). Isotope effect in the Solubility of Hydrogen in FCC Metals. *Physica Status Solidi (a)*, Vol. 9, Iss. 1, (January 1972), pp. 289-300, ISSN-1862-6319

Zinkle, S. J. & Fabritsiev S. A. (1994). Copper alloys for high heat flux application. *Atomic and Plasma-Material Interaction Data for Fusion*. Vol 5 (December 1994), pp. 163-191.

Part 2

Development of High-Performance Current Copper Alloys

3

Properties of High Performance Alloys for Electromechanical Connectors

H.-A. Kuhn, I. Altenberger,
A. Käufler, H. Hölzl and M. Fünfer
Wieland-Werke AG, Ulm,
Germany

1. Introduction

Miniaturization of electronic and electromechanical components and increasing cost of materials are the driving forces for developments of high performance copper alloys for automotive and computer technologies. Electromechanical connectors are current carrying spring elements. Miniaturization requires improved mechanical strength and medium to high electrical conductivities. For automotives this components have to operate in temperature ranges between -40°C and 180°C under hood. For a good reliability during life-time of vehicles or multimedia devices designing engineers also expect good formability, excellent resistance against stress relaxation and fatigue behavior. The design of small box-like connectors has to overcome the contradiction between high strength and good formability. Nature of alloys also reveals reduced electrical and thermal conductivity at improved strength and vice versa reduced strength at higher conductivity. Nevertheless, development of modern copper based connector materials intends to improve strength, electrical conductivity and bending behavior simultaneously. Furthermore, a long life time of connector devices requires excellent thermal stability and resistance against fatigue damage.

In a first step the following contribution will focus on the metallurgical principles for alloy design. The role of alloying elements is described. The diverse directions of material development include optimization of standardized alloys like tin bronzes and development and processing of new alloy compositions. Beside properties of grain refined tin bronze CuSn8 with standardized chemical composition this article presents latest developments on precipitation hardened Corson-type alloys.

Corson has first described in 1927 the mechanism of precipitation of nickel silicides in copper (Corson, 1927). The successful story of CuNiSi-alloys for connector devices has started in the early 80ies (Hutchinson, 1990; Tyler, 1990; Robinson, 1990). Until today research is focused on understanding processing and optimization of standardized C70250 (UNS-designation) like CuNi2Si and related alloys of the first generation (Lockyer, 1994; Kinder, 2009). The success of an ongoing research on these alloys is due to an excellent combination of strength, conductivity, deformation behavior and thermal stability. Therefore these alloys have been applied to diverse connector and lead frame devices. In addition, many connector devices are subjected to dynamical load (vibrations under hood) or to high numbers of repeating plug ins resulting in fatigue damage of spring elements.

Since a few years this materials are followed by more complex cobalt containing Corson-type alloys CuNi1Co1Si (C70350) which stand comparison with some copper-beryllium alloys of medium yield strength up to 850 MPa (Mandigo, 2004; Kuhn, 2007; Robinson, 2008). High strength copper nickel silicon alloys also meet requirements of high density processor sockets (Robinson, 2008). So called Hyper Corson-type alloys with more than 5 wt% of precipitates forming elements achieve 900 MPa and more (Mutschler, 2009).

This article is focused on properties of spring behavior with concern to elasticity, stress relaxation resistance and fatigue behavior with respect to microstructures of fine grained tin bronzes and precipitation hardened Corson type alloys. Deformation behavior and bendability of strip materials were already reported by (Kuhn, 2007; Bubeck, 2007).

2. Experimental procedure

2.1 Characterization of microstructure

For optical light microscopy grain boundaries, grain sizes and second phases of thermo-mechanical treated rolled strips were revealed by immersion etching in sulphuric acid solution of $K_2Cr_2O_7$ (Hofmann, 2005).

Silicides of Corson-type alloys with diameters less than 200 nm were characterized by a Scanning Electron Microscope SEM (Zeiss Ultra). By SEM using secondary electrons as well as backscattered electrons at magnifications up to 20.000 x it was possible to image small precipitates with radii of 20 nm and dislocation induced deformation areas. Semi quantitative chemical analysis of particles was performed by Electron Dispersion X-Ray Spectroscopy EDXS (ISIS300, Oxford Instruments).

Precipitates of the Corson-type alloy C70350 were also investigated by transmission electron microscopy (JEOL 4000 FX, 400 kV). Investigations were conducted by Stuttgart Center for Electron Microscopy.

2.2 Mechanical properties

Strength is measured as Yield Strength $R_{p0,2}$ and Ultimate Tensile Strength R_m. Strengths and elongations of strips were examined at room temperature by a tensile tester Z100 (Zwick).

2.3 Modulus of resilience

The modulus of resilience is defined by the stored energy of a spring (Fig.1):

$$\text{Resilience} = R_{p0,2}^2/2E \qquad (1)$$

The calculation of the modulus of resilience requires exact values of the yield strength $R_{p0,2}$ and the Young's modulus E. For a rough estimate the Young's modulus obtained from the classical tensile test is usually used. For more precise determination of the modulus of resilience the Young`s modulus is preferably determined by the dynamical resonance method (Förster, 1937).

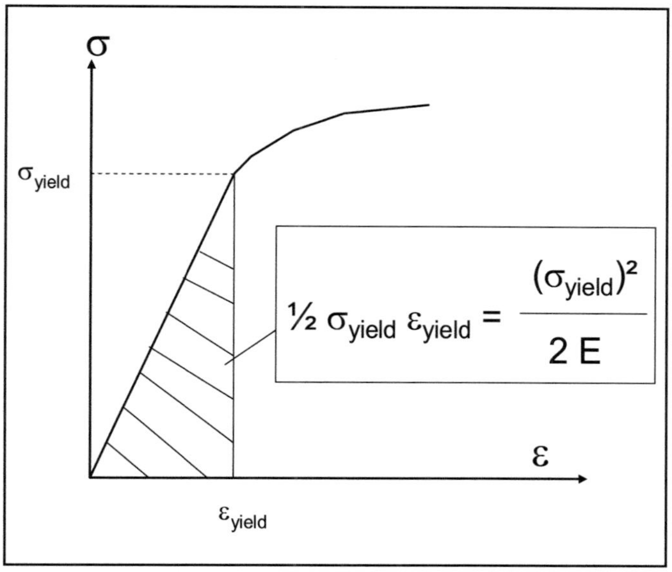

Fig. 1. Definition of the modulus of resilience of a metallic material used as a spring.

Young´s modulus E was determined by the dynamical resonance method as described in (Kuhn, 2000). E was evaluated in dependence on the resonance frequency f of the 1st transversal vibration, the specific weight ρ , length l and thickness t of the tested strip materials. The correction factor C is a function of Possion´s ratio and geometry. For resonance frequencies of 0.5 to 1 kHz specimen sizes of t x 20mm x 50mm were chosen. Thickness varies between 0.15 and 0.5 mm depending on the manufactured final strip gauges.

$$E = 0.94645\rho f^2 \,(l^4 / t^2)\, C \qquad (2)$$

2.4 Stress relaxation resistance

Thermal stability is expressed by the material's resistance against stress relaxation (SRR) under a defined load at elevated temperatures between 80 and 200°C after 1000 to 3000 hours. For the tried and tested ring method (Fox, 1964) strips of 50 mm length and gauges between 0.15 and 0.5 mm were clamped on a steel ring with a defined outer radius. The applied initial stress σ_i in the outer surface of the strip depends on this radius and the yield strength of the copper alloy. For σ_i usually 0.5 to 1.0 x $R_{p0,2}$ were chosen. By the time under load and temperature elastic stress changes into plastic deformation. The remaining stress is an indicator for SRR. Measured data were extrapolated to long life times by the Larson-Miller approach (Larson & Miller, 1952). Test equipment and evaluation of results is well described by (Bögel, 1994; Bohsmann, 2008).

2.5 Fatigue behavior

All fatigue tests were conducted under air atmosphere at 22°C. Specimens made from 0.3-0,6 mm strip material where cycled by alternating bending (R-ratio R = -1) using a mechanical testing device at a frequency of 18 Hz under constant extension amplitude. The stress amplitude is calculated through the elastic properties (Young's modulus and geometry) of the cycled specimen with rectangular cross section. The rectangular specimens were loaded perpendicular to the rolling direction of the strip. SEM pictures were taken using a ZEISS Ultra Scanning electron microscope (equipped with a Schottky field emission cathode). Roughness values in axial direction prior to fatigue testing were R_a = 0,08 µm and R_z = 0,84 µm for the coarse grained CuSn8 and R_a = 0,11µm and R_z = 0,78 µm for the very fine grained CuSn8. R_a and R_z of the investigated CuNi1Co1Si-Strip were 0,39 and 0,83 , respectively.

3. Alloys

Designer of electromechanical connectors first have to decide for high strength or high conductivity. Connectors under high current load need copper alloys with medium to high electrical conductivity. For transmission of signals copper alloys with low to medium conductivity are used. It is known that copper alloys of high strength cannot reveal high conductivities like pure or low alloyed solid solution hardened coppers. The chart of electrical conductivity versus yield strength (Fig. 2) gives a first overview about alloying systems of relevance. In case of precipitation hardened low alloyed coppers, Cu-Be-alloys and Corson-type alloys this two-dimensional interdependence is valid for good to fair bend radii r versus strip thicknesses t ratios r:t in the range of 0 to 5.

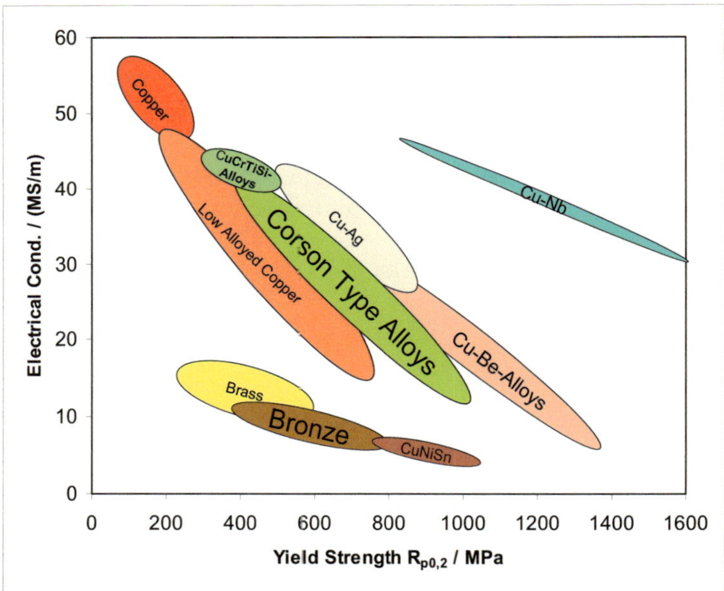

Fig. 2. Copper and Copper alloys: Electrical conductivity versus yield strength.

3.1 CuSn8

In principle, the development of copper alloys for connectors is not only driven by improvement of their properties. It is also restricted by demands from connector manufacturers for standardized materials, double sourcing and materials costs. By introducing very fine grained strip materials of standardized alloys into the connector market this contradiction in customer demands can be overcome. Phosphor containing bronze CuSn8 (European Standard: CW453K; UNS-designation: C52100) strip materials with an average grain size of less than 2 µm reveal improved strength and improved formability in comparison with materials of usually medium grain sizes of 10 to 20 µm (as shown in (Bubeck, 2007; Bubeck, 2008)). This grain refinement increases yield strength in the order of 100 to 150 MPa without any loss in formability. This is a result of optimized thermo-mechanical treatment of rolled strips. Chemical composition was not varied. Time and temperature of intermediate recrystallization annealing steps were fit to defined amounts of cold reduction in order to suppress grain growth. Grain size decreases after each cold roll and recrystallization annealing step until an average grain size of 1 to 2 µm is achieved prior to final cold rolling. Hardening in this class of alloys is based on the Hall-Petch relation (Fig. 3) (Bubeck, 2007; Bubeck, 2008; Hall, 1951). Grain refinement leads to improved strength and ductility of strip material in comparison with conventionally processed CuSn8. This advantageous combination of mechanical properties enables manufacturers to design small box like connectors of medium to high strength with nearly sharp bend edges. Improvement of bending behavior was discussed earlier in detail (Bubeck, 2007; Bubeck, 2008). Mechanical properties of investigated strip materials are listed in Table 1.

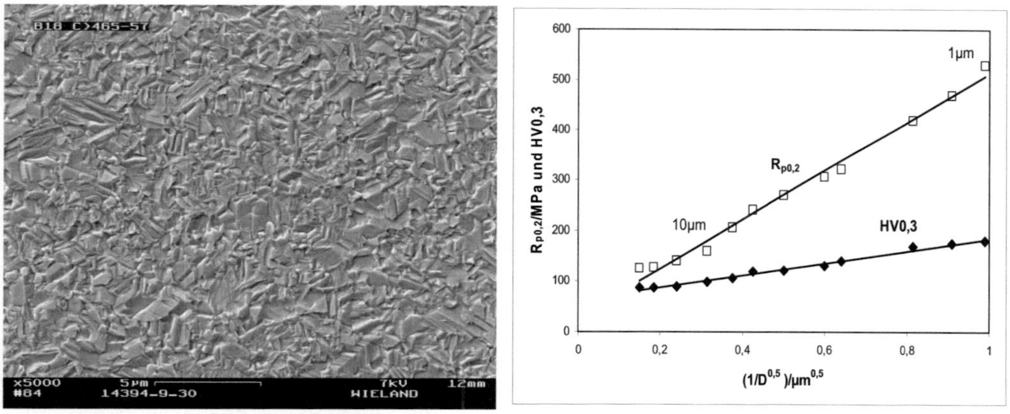

Fig. 3. CuSn8 strip with grain size 1 to 2 µm (a), Hall-Petch-Relation: Hardness and Yield strength vs grain size (Bubeck, 2007) (b) .

Alloy	$R_{p0,2}$/ MPa	R_m / MPa	A10 / %	Electr. Conduct. / MS/m	Minimum bend radius r:t GW/ BW*
CuSn8 Grain size 10 µm	610	665	33	8	0,5 / 1,5
CuSn8 Grain size 1..2 µm	650 ...740 780 ...830	685 ...785 800 ...925	15 ...20 5 ... 8	8 8	0 / 2 1 / 4
CuNi3Si1Mg	725	770	10,4	23,5	1,5/1,5
CuNi1Co1Si TM04	778	816	11,4	31	2,0 / 2,0
CuNi1Co1Si TM06	825	860	5,5	28	2,5 / 2,5
CoNi3,9Co0,9Si1,2Mg0,1	870 993	930 1018	2,0 2,1	18 18	2,1 / 2,1 4,4 / 7,2

* GW: = Bend edge in good way direction (perpendicular to rolling direction), BW: = Bend edge in bad way direction (parallel to rolling direction).

Table 1. Mechanical properties, electrical conductivity and minimum 90° bend radii

3.2 Corson type alloys

The more traditional development approach for high strength copper alloys is focused on chemical variation of precipitation hardened alloys.

As earlier explained by (Hutchinson, 1990), silicide-containing Corson-type alloys CuNi3Si or CuNi2SiMg represent a class of materials with high strength copper materials beside the superior beryllide hardened CuBe-alloys. Processing of beryllium-containing particles during production and manufacturing of beryllium containing metals can cause lung disease (Schuler, 2005). Be–free compositions are legislatively urged in the European Union recently. Therefore many manufacturers and consumers prefer Be-free copper alloys and further developments of silicide hardened copper alloys aim at replacing copper-beryllium alloys. Targets are combinations of yield strength and electrical conductivity of 800 MPa and 50% IACS or 950 MPa and 30% IACS. For comparison the first generation of Corson alloys used for connector materials like CuNi2Si, CuNi2SiMg and CuNi3Si with volume fractions of precipitates of 2 to 3 percent is characterized by 700 MPa and 50% IACS. Beside these fundamental properties new materials have to meet requirements for excellent thermal stability with 70% remaining stress after 1000h at 150 °C and good formability.

Variations of the chemical compositions of basic Corson-alloys aim to increase the volume fraction of particles and/or improve thermal stability of precipitates. For this purpose one has to understand the role of alloying elements.

Nickel reduces the solubility of silicon in copper. This is also valid for cobalt. From the quasi binary phase diagrams, one can deduce an increase of solvus of silicides in dependence of the sum of the silicide forming elements Ni, Co and Si. The temperature of solvus line increases also with the ratio of Co versus Ni.

For example the solvus temperature of pure Ni_2Si is 770 °C for 2.5 wt% of Ni plus Si and 800 °C for 3 wt%. For comparison the solution annealing temperature of pure Co_2Si is approximately 1070 °C for 2.5 wt% Co plus Si. With a Ni:Co ratio of 1 at composition sum of Ni , Co and Si of 2.5 wt% the solution annealing temperature is reduced to 960 °C. These data are taken from (Mandigo, 2004). The superior hardening effect of Co-silicides was also demonstrated by (Fujiwara, 2004). The strengthening effect depends on the size, crystallographic structure and the distance between the involved particles. The radius of effective silicides is less than 100 nm.

A prerequisite for a successful solution annealing is subsequent quenching of annealed strips. For solution annealing strips, typical solution annealing treatments take place at 900 to 1000 °C. Optimized solution annealing can be verified by the minimum of electrical conductivity. For further processing of strips, grain sizes of 20 µm and less are desirable.

In the following the quenched strips are age hardened at 400 to 500 °C. With respect to isotropy of bending the amount of final cold roll prior to final age hardening is usually 25 to 50 %. An optimized combination of yield strength and electrical conductivity can be obtained when the (Ni+Co)/Si ratio is between 3.8 and 5.

Due to the enhanced solution temperature, copper alloys hardened by cobalt-silicides Co_2Si reveal improved stress relaxation resistance as compared to alloys containing only Ni_2Si.

Non shearable precipitates cause additional strength in dependence of their volume fraction and their radii. Strengthening of Copper-alloys hardened by silicides is based on the

Orowan–mechanism due to the semi-coherency of the orthorhombic crystal structure oP12 of $(Ni, Co)_2Si$ precipitates (Kuhn, 2007). The contribution of particle hardening $\Delta\sigma$ to yield strength can be improved by an increase of volume fraction f and a reduction of particle radius r of precipitates:

$$\Delta\sigma \cong 6\frac{\sqrt{f}}{r} \tag{3}$$

This formula was derived from the Orowan approach (Orowan, 1954)

$$\Delta\sigma \cong 0.8Gb\sqrt{n_A} \tag{4}$$

(G: = shear modulus of copper; b: = Burgers-vector; n_A:= number of particles per unit area within the slip plane)

Fig. 4 explains the effect of particle size and volume fraction on the strengthening by non-shearable particles. For a gain of 100 MPa in yield strength a microstructure with a majority of very small precipitates with radii less than 10nm and at a volume fraction of 3 and more percent is required.

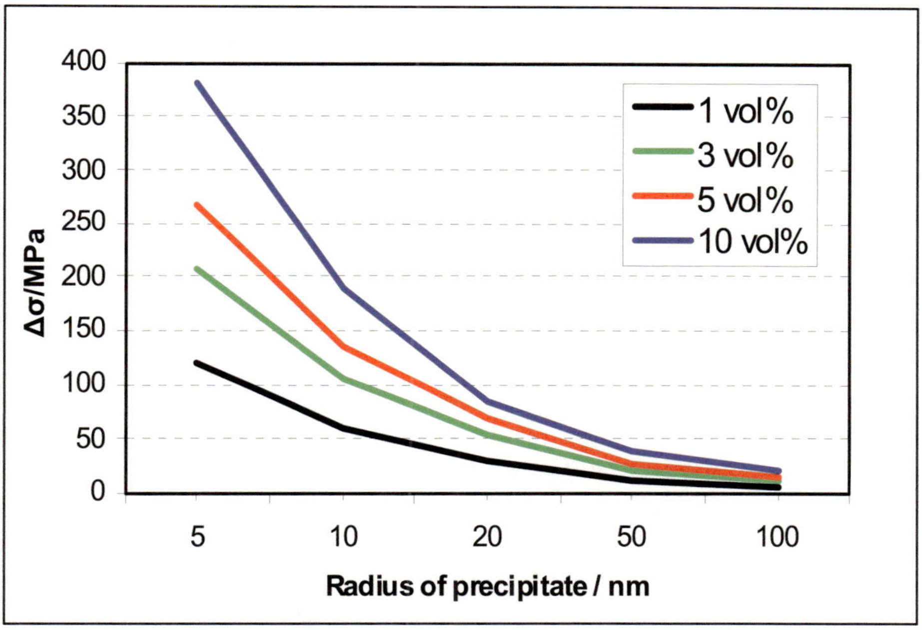

Fig. 4. Effect of particle size and volume fraction on yield strength increase according to the Orowan-mechanism.

Fig. 5 shows typical precipitate distributions of $(Ni, Co)_2Si$ in CuNi1Co1Si strip (left micrograph) and of CuNi3SiMg (right micrograph). Coherency strains as figured out by the following TEM-micrograph indicate particle diameters less than 5 nm (Fig.6) in CuNi1Co1Si. Obviously the radii of most particles vary between 2 and 100 nm.

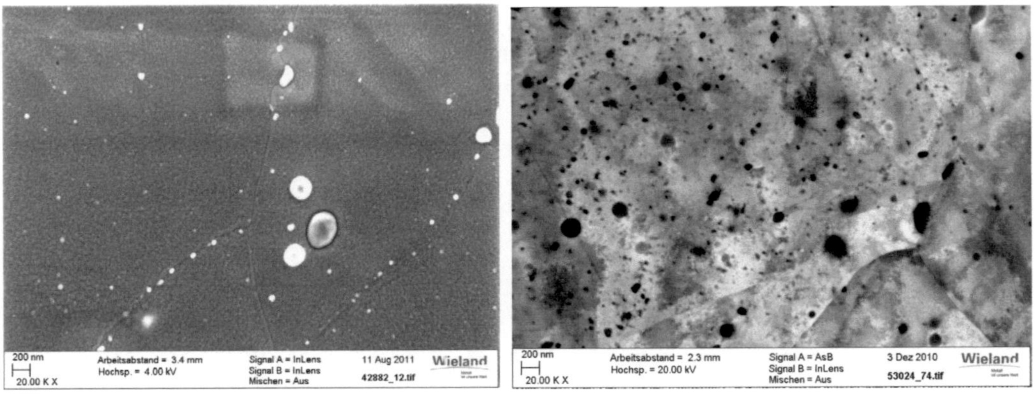

Fig. 5. SEM-micrographs: precipitates in CuNi1Co1Si, secondary electron image (left) and CuNi3SiMg, backscatter electron image (right).

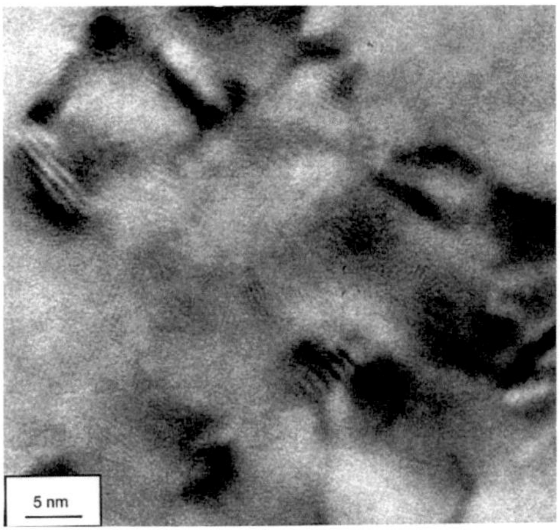

Fig. 6. TEM-micrograph of CuNi1Co1Si: Small particles and coherency strains (by courtesy of Stuttgart Center of Electron Microscopy).

Beside cobalt, nickel and silicon, other elements such as magnesium and chromium contribute to the improvement of strength and thermal stability of Corson-type alloys. Unlike in the microstructure of lower alloyed CuNi1Co1Si chromium forms coarse chromium containing silicides. One disadvantage of such coarse precipitates is the wear of stamping tools in the connector production root. Magnesium atoms improve stress relaxation resistance by solid solution hardening due to their large difference of atom radii r_{Mg} in comparison with copper r_{cu}. $(r_{Mg}-r_{Cu}/r_{Cu})$ is of the order of 25 %. Mg also contributes to the formation of silicides as known from the C70250 variation CuNi3SiMg.

In the following, the latest developed Corson-type alloy CuNi3,9Co0,9Si1,2Mg0,1 with yield strength beyond 900 MPa characterized by Ni and Co-containing mixed silicides and a composition of more than 5 wt % alloying elements is called Hyper Corson alloy. Unlike the microstructure of lower alloyed C70250 and C70350 some silicides of Hyper Corson alloys exhibit sizes of 5 µm. However large silicides do not contribute significantly to strengthening, they inhibit grain growth.

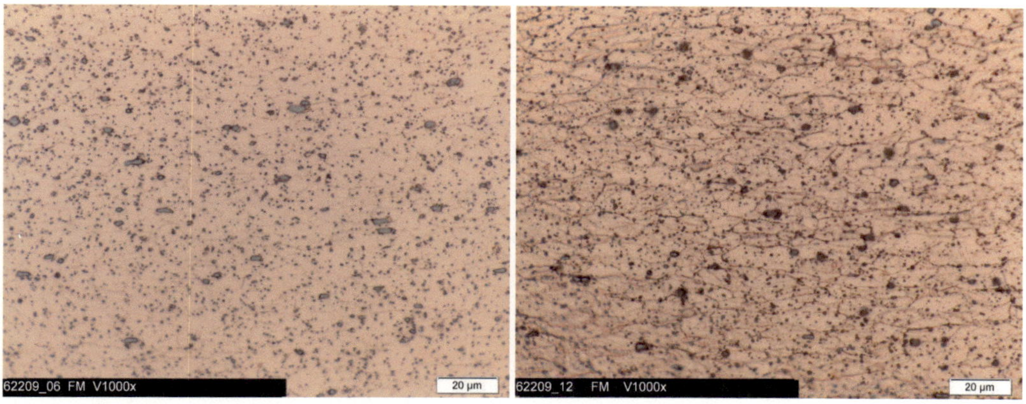

Fig. 7. Hyper Corson alloy CuNi3,9Co0,9Si1,2Mg0,1: Distribution of silicides after age hardening. Etched micrograph (right figure) indicates grain sizes of 10 to 20 µm in rolling direction.

4. Results

4.1 Stress relaxation resistance (SRR)

Mechanical properties and electrical conductivity of investigated strips of Corson type alloys are summarized in table 1.

For precipitation hardened high strength Corson type alloys used for automotive devices the acceptable SRR is at least 70% of initial stress remaining after samples are exposed for 3000 hours at 150°C. The Mg-containing alloy CuNi2SiMg (C70250) as an example for the first generation of Corson type alloys used for connector materials meets this target at an initial stress in the magnitude of the yield strength as well as the Co-containing second generation CuNi1Co1Si (C70350). The initial stresses are 738 MPa for CuNi3SiMg, 751 MPa and 739 MPa for CuNi1Co1Si TM04 and TM06 , respectively, and 670 MPa for CuNi3,9Co0,9Si1,2Mg0,1. As shown in Fig. 8 for 150°C with initial stresses of 70% or equivalent to the yield strength the SRR of C70250 expressed by the remaining stress after load time is higher compared to C70350 and the Hyper Corson alloy.

Between 1000 and 3000 hours of load the average loss of stress indicates a time dependence of 12 kPa/hour for C70250 and 7 kPa/hour for both C70350 tempers. This is a first hint for an improved SRR at higher temperatures for C70350 and Hyper Corson alloy. The advantage of the Co-containing alloy becomes more obvious at 200°C. Both temper designations of C70350 and the Hyper Corson alloy exhibit a strong SRR beyond 1000 hours of load compared to R690 of CuNi3SiMg (Fig. 8). An extrapolation to 10.000 hours forecast the total loss of initial stress for CuNi3SiMg, whereas for both CuNi1Co1Mg tempers stresses of 40 to 48 % of initial strength will remain. This is also valid for the Hyper Corson alloy with a remaining stress of 55% of initial strength. For example, between 1000 and 3000 hours time of load the averaged time dependent loss of initial stress is 52 kPa/h for CuNi3SiMg and only 7 kPa/h for all Co-containing alloys.

Advantages of Co-containing Corson alloy are revealed above 150 °C. One of the reasons for this behavior is the thermal stability of Co-containing silicides as predicted by binary phase diagrams. In relation with necessary solution anneal temperature of pure Co_2Si an elevated temperature of 473 K (200°C) is 0,37 times of T_{sol} compared to 0,42 of Ni_2Si. More decisive is the distance in temperatures between solvus line and age hardening temperature with respect to the amount of cold roll prior age hardening. In case of Co-containing Corson alloys a larger number of small silicides is expected due to a very high solution anneal temperature.

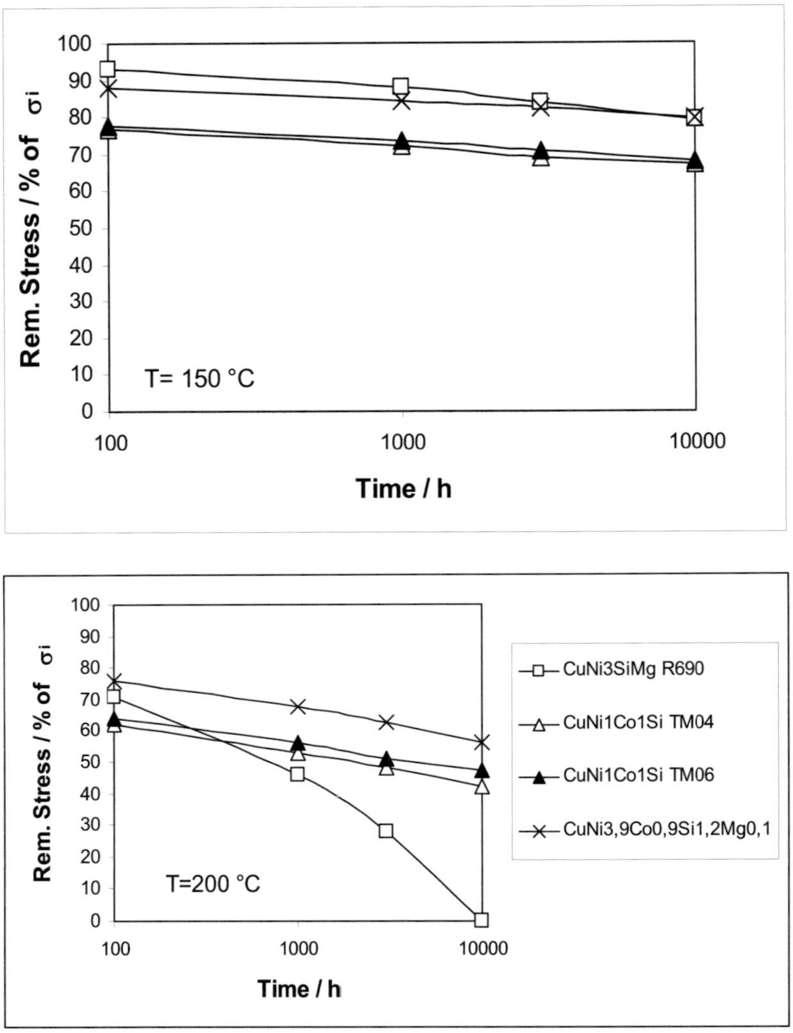

Fig. 8. Comparison of resistances against stress relaxation for temperatures 150 °C and 200 °C.

4.2 Fatigue behaviour

One of the most important properties for the functionality of copper-based connectors and springs is the ability to withstand repeated cyclic loads for many million cycles under bending-, push-pull or torsional loading conditions. The fatigue strengths of high-performance copper alloys exceed those of pure copper by up to several hundred MPa and are superior to those of non-hardened plain carbon steel, austenitic steels or at least equal to quenched and tempered steels [1]. Fig. 9 gives an overview of the fatigue strength (endurance strength for 10^7 cycles) of several copper alloys.

Similar to other metallic materials and alloys, there is (for smooth, unnotched specimens) an empirical relationship between the fatigue strength and the tensile strength. Within the same chemical composition the fatigue strength rises with increasing tensile strength. The fatigue strength/tensile strength ratio strongly differs from alloy to alloy and is in the range of 0.25-0.55. The fatigue strengths of electromechanical connectors alloys such as CuSn8 and CuNi1Co1Si under reversed bending are in the range 300 - 450 MPa.

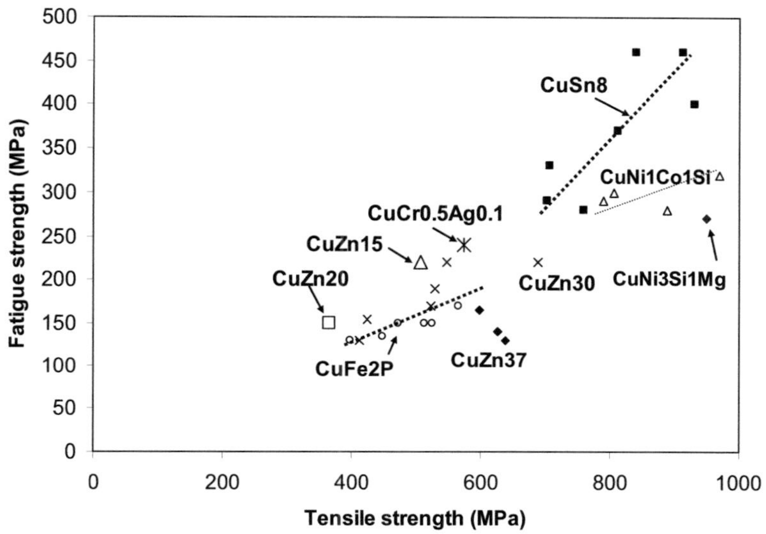

Fig. 9. Fatigue strengths of different copper-based alloys (fully reversed bending (R = -1), samples: copper alloy strip of 0.3 to 0.6 mm thickness, loading transversal to rolling direction)

4.2.1 Fatigue of tin bronzes

Although non-precipitation hardened single phased α-alloys (such as tin bronzes, e.g. CuSn8) are inferior to precipitation hardened copper alloys (such as CuNi1Co1Si) in respect to thermal stability of microstructure and stress relaxation behavior, they exhibit excellent fatigue strengths if they are sufficiently cold worked or hardened by grain refinement, (either by thermomechanical processing (Bubeck, 2007; Bubeck, 2008) or by severe plastic deformation). Fig. 10 shows the statistically evaluated stress-life (S/N) behaviour of CuSn8 in two different grain sizes, namely for average grain size of 20 μm and for an average grain size of 1,5 μm. It can be seen clearly, that the very-fine-grained condition shows a significantly higher fatigue life and -strength in the investigated Low Cycle Fatigue (LCF) as well as in the High Cycle Fatigue (HCF) regime.

Further fatigue strength enhancement, especially in the HCF-regime, can be achieved by ultrafine grained (UFG) or nanocrystalline material (Höppel, 2002). Grain refinement as a hardening mechanism is especially attractive in maintaining sufficient ductility for cold

[1] In the Very-High-Cycle (VHCF) fatigue regime (10^8-10^{11} cycles to failure) the endurance strength of pure copper is clearly below 100 MPa (Weidner, 2010; Heikkinen, 2010)

forming processes. Very fine grained tin-bronze is thus characterized by an excellent combination of ductility and strength (Bubeck, 2007; Bubeck, 2008).

Typical crack initiation sites in high performance copper alloys are similar as in pure copper qualities, namely persistent slip bands (PSBs) at medium or high stress amplitudes, twin boundaries at low stress amplitudes [26] or grain boundaries or grain boundary triple points at high stress amplitudes. In addition, dislocation pile ups at precipitates or non-metallic inclusions may play a damaging role, especially in the VHCF-regime, below the threshold of PSB-formation.

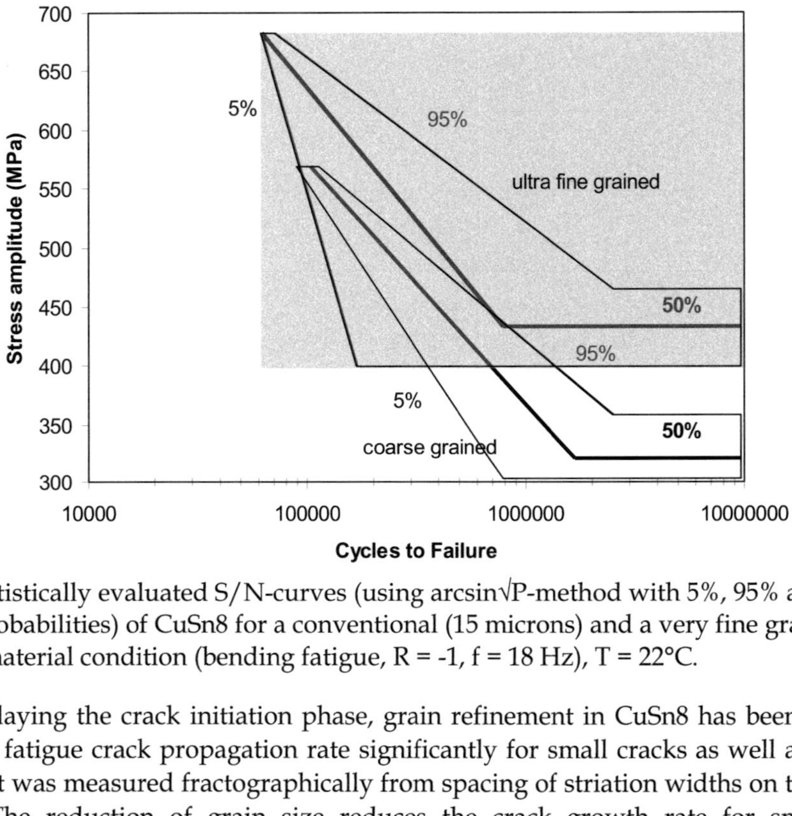

Fig. 10. Statistically evaluated S/N-curves (using arcsin√P-method with 5%, 95% and 50% fracture probabilities) of CuSn8 for a conventional (15 microns) and a very fine grained (1-2 microns) material condition (bending fatigue, R = -1, f = 18 Hz), T = 22°C.

Besides delaying the crack initiation phase, grain refinement in CuSn8 has been shown to reduce the fatigue crack propagation rate significantly for small cracks as well as for large cracks, as it was measured fractographically from spacing of striation widths on the fracture surfaces. The reduction of grain size reduces the crack growth rate for small cracks according to the slip band model (Tanaka, 2002). The reduced cyclic plasticity at the crack tip of very fine grained CuSn8 is reflected in different morphologies of the fracture surfaces of conventional and very fine grained CuSn8 (Fig. 11). The fine grained structure shows a rather smooth fracture surface with little crack tip blunting (see e.g. model by (Laird, 1962)) and less pronounced striations as in the coarser material condition.

The role of twinning for mechanical and thermal properties of high performance copper alloys remains controversial. On the one hand twin boundaries are known to be one of the preferential regions for fatigue crack initiation especially during High-Cycle-Fatigue (Bayerlein, 1991). On the other hand they can enhance the thermal stability of ultrafine grained (UFG) materials (Anderoglu, 2008) and act as weak obstacles to dislocations (Qu & Zhang, 2008). Moreover, deformation twinning can act as a supplementary mechanism to enhance ductility, formability and bendability. The density of twins can be controlled by adjusting the grain size and the stacking fault energy of the alloy: The density of deformation twins decreases with decreasing grain size and the (largely grain-size insensitive) density of recrystallization twins increases with decreasing stacking fault energy (Vöhringer, 1972; Vöhringer, 1976).

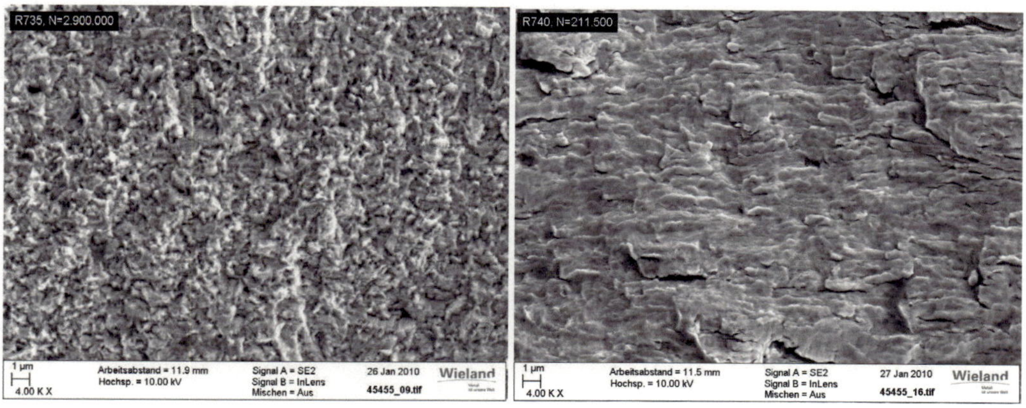

Fig. 11. SEM-micrograph of the fracture surface of fatigued CuSn8 (left: very fine-grained, right: coarse-grained condition)

Fig. 12 shows an Orientation Imaging Microscopy (OIM) picture obtained by Electron Backscatter Diffraction (EBSD) of conventional and very fine grained CuSn8. The coarse-grained CuSn8 exhibits significantly more twin boundaries (characterized by a 60°-misorientation between adjacent grains or a coincident site lattice (CSL) number of $\Sigma = 3$) than the fine grained conditions (compare Figs. 13 and 2). The optimization of the nature, density and location of twins for enhanced mechanical behavior of high performance copper alloys remains a challenging task within grain boundary engineering (Randle, 1999).

Fig. 12. OIM (Orientation Imaging Microscopy)-picture (obtained by EBSD) of coarse grained (right) and fine grained CuSn8 (left) prior to final cold rolling. Cu-type orientation is represented by blue coloured grains, Goss-type by green coloured grains[2]

[2] Calculations of Schmid- and Taylor-factors gave similar factors for both fine- and coarse-grained conditions in the order of ~2.8

4.2.2 Fatigue of precipitation hardened alloys

A combination of high fatigue strength and excellent stress relaxation stability and thermal stability can be achieved with precipitation hardened copper alloys such as CuNi1Co1Si (see also Fig. 9) or CuNi3Si1Mg. Optimized precipitation hardening promotes fatigue strength as well as high electrical conductivity. The high fatigue strength in Corson- or Corson-derivated alloys is principally derived from Ni- or Co-Silicides (Lockyer, 1994) as well as Ni-Co mixed silicides which are orthorhombic and semicoherent with the α-Matrix and serve as dislocation obstacles according to the well-known Orowan-mechanism (Orowan, 1933; Ardell, 1985). Fig. 13 shows precipitates in the fracture surface of fatigued CuNi1CoSi.

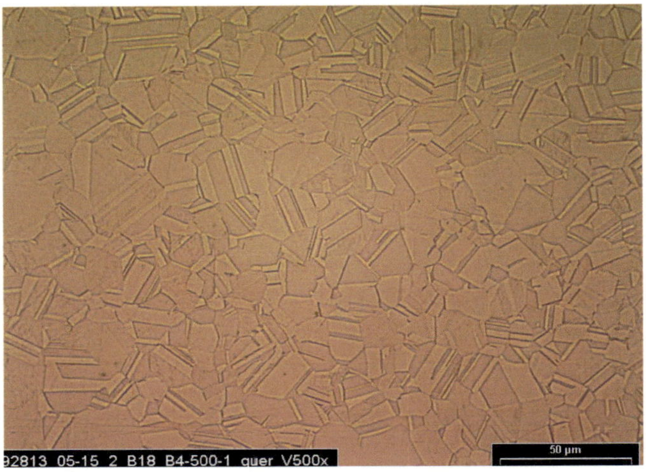

Fig. 13. Microstructure of conventionally processed CuSn8 with 2-4 twins per grain.

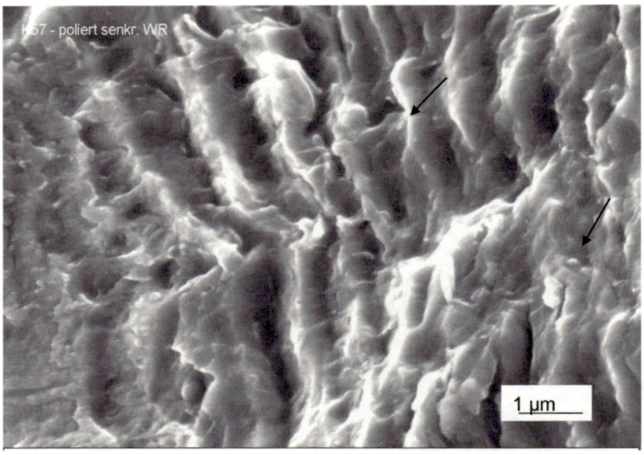

Fig. 14. Fatigue fracture surface of a spring (material: CuNi1Co1Si) exhibiting striations and precipitates (arrows).

The slopes of the fatigue strength dependence of the tensile strength as presented in Fig. 9 differ between solid solution hardened one phase CuSn8 and all precipitation hardened Corson-type alloys. The flat slope of precipitation hardened alloys can be explained by the interaction between moving dislocations and non-shearable semi coherent particles. This mechanism may cause micro cracks around the second phases as shown in fig. 15 for CuNi1Co1Si. Due to differences in Young´s modulus (E_{Matrix} : ~130 GPa, $E_{Silicide}$: ~160 GPa (Lockyer, 1995)) additional localized strain gradients were generated. Elastically deformed particles generate residual stresses and geometrically necessary dislocations.

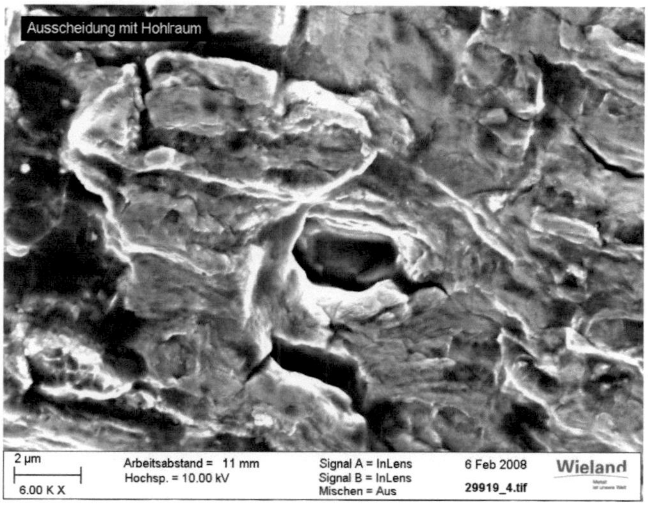

Fig. 15. Fatigued strip of CuNi1Co1Si TM04: Formation of cracks in the interface between matrix and (Ni,Co)-silicide.

4.3 Modulus of resilience

The elastic behaviour of electromechanical connector materials is of utmost importance since they, besides conducting electric current, act as springs. In order to maintain their grip during service the connectors have to excert high elastic forces onto their elastic contact partner, ideally without any time- or temperature-dependent stress relaxation.

The ability of a component to store elastic energy is described by the *modulus* of *resilience*. Springs with a high resilience can store and release (upon unloading) a lot of elastic energy, springs with a low resilience can store and release only little elastic energy. In a stress-strain-plot the modulus of resilience of any material is defined as the area under the stress-strain curve (or *strain energy density*) up to the elastic limit (yield strength) of the material (Fig. 1). The modulus of resilience rises with increasing yield strength and with decreasing Young's modulus.

Fig.16 shows the modulus of resilience for several copper alloys including electromechanical connector alloys CuNi3SiMg and CuNi1Co1Si. The modulus of resilience of electromechanical connector alloys is in the range of 0.8-3.5 MPa (depending on the strength and temper condition, see also (Kuhn, 2007). Finally, in respect to resilience only, a perfect copper-based spring would be made from metallic glass (such as $Cu_{47}Ti_{34}Zr_{11}Ni_8$). Unfortunately, the electrical and thermal conductivity of bulk metallic glasses are rather low. Conversely, composites consisting of metallic glass and a pure copper or copper alloy with high conductivity could offer interesting solutions for high strength-high conductivity applications (Choi-Yim, 1998).

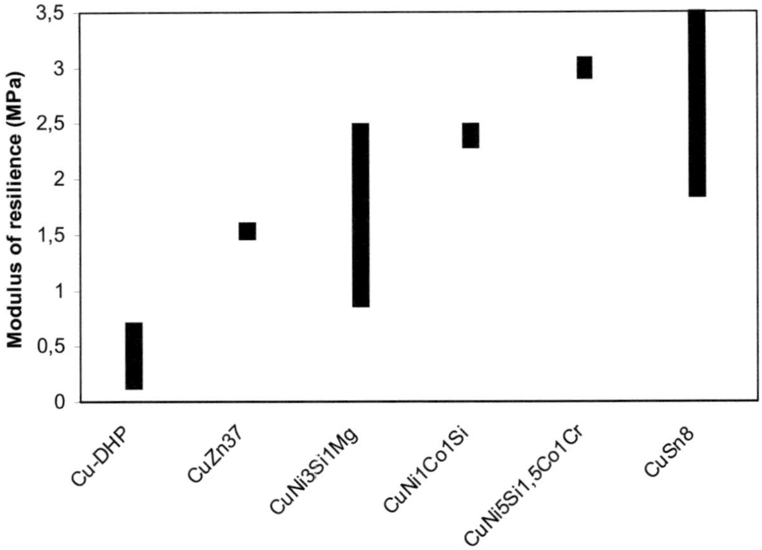

Fig. 16. Modulus of resilience for several copper alloys.

It should be noted that the modulus of resilience is temperature- and orientation- and therefore texture-dependent. In strips exhibiting a typical rolling texture, a high modulus of resilience can be found perpendicular and parallel to the rolling direction, whereas in angles in between these orientations lower values for the modulus of resilience are observed (Fig. 17).

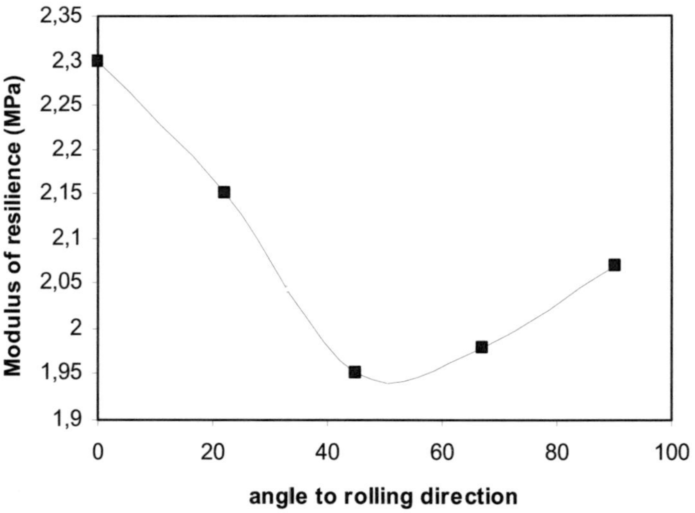

Fig. 17. Modulus of resilience of CuNi3Si1Mg in different angles to the rolling direction of a strip.

5. Conclusions

Modern Cu-based electromechanical connector alloys are characterized by a significant improvement of stress relaxation resistance in the temperature range 150°C - 200°C as compared to classical Corson-type alloys (containing only nickel-silicides) of the first generation. The main microstructural reason for this superiority is the precipitation of finely dispersed mixed Co-Ni-silicides. Due to the high undercooling during quenching of the solution treated condition the nucleation of nanoscale $(Co,Ni)_2Si$ is enhanced. Transmission electron microscopy studies revealed precipitate diameters of < 10 nm which is assumed to be in the order of the maximum hardening contribution by precipitates. A statistical investigation of precipitate sizes and their correlation to Co-content and thermomechanical process steps therefore seem to be promising to further evaluate the potential of the Cu-Ni-Si-Co system.

High alloyed Hyper-Corson alloys containing alloying elements > 5 mass% exhibit excellent mechanical properties and therefore appear to be an attractive class of materials to replace Cu-Be alloys.

In contrast to precipitation hardened copper alloys of similar strength, single-phased fine grained and solid solution hardened tin bronzes are able to withstand higher cyclic loads in fatigue tests. It is assumed that the less pronounced strength-dependency of the fatigue strength of precipitation hardened CuNiSi alloys can be attributed to internal stress raisers in the microstructure, e.g. generation of micro cracks on the phase boundary between precipitate and matrix due to deformation incompatibilities and increased stress by pile ups of dislocations.

Fatigue investigations have revealed that significant grain refinement of tin bronze leads to superb fatigue strength. The grain refinement goes hand in hand with a reduction of twin density, which on the other hand reduces the number of possible crack initiation sites.

For spring applications, a high modulus of resilience is predominantly achieved by high yield strength. High strength Co-containing Corson-type alloys can therefore compete with CuSn8 inspite of lower Young's modulus.

6. References

Anderoglu, O., Misra, A., Wang, H., Zhang, X., (2008). Thermal stability of sputtered Cu films with nanoscale growth twins: *Journal of Applied Physics*, Vol. 103.

Ardell, A.J., (1985). Preciptiation hardening: *Metallurgical Transactions A*, Vol. 16, p. 2163.

Bayerlein, M., (1991). Doctorate Thesis, Universität Erlangen-Nürnberg.

Bohsmann, M., Gross, S., (2008). Understanding Stress Relaxation in: *Proc. Materials Science & Technology (MS&T)*, Pittsburgh, Pennsylvania, p. 41.

Bögel, A., (1994). Spannungsrelaxation in Kupferlegierungen für Steckverbinder und Federelemente, *Metall*, Vol. 48, p. 872.

Bubeck, F., Kuhn, H.-A., Buresch, I., (2007) *Mischkristallgehärtete Kupferbasislegierungen für Steckverbinder Optimierung der Eigenschaften in:* 14. Sächsische Fachtagung für Umformtechnik, Freiberg 4. Dez. 2007, (edit. R. Kawalla).

Bubeck, F., Gross, S., Buresch, I., (2008). *SUPRALLOY: A new High Performance Bronze* in: Proc. Materials Science & Technology (MS & T), Pittsburgh, Pennsylvania, p. 32.

Choi-Yim, H., Busch, R., Johnson, W.L., (1998). *Journal of Applied Physics*, Vol. 83, p. 12.

Corson, M.G. (1927). Copper Hardened by a New Method, *Zeitschrift für Metallkunde*, Vol. 19, p. 370.

Fox, A., (1964). *Research and standards*, Vol. 4, p. 480.

Förster, F., (1937). Ein neues Messverfahren zur Bestimmmung des Elastizitätsmoduls und der Dämpfung, *Zeitschrift für Metallkunde*, Vol. 29, p. 109- 113.

Fujiwara, H., (2004). Designing High-Strength Copper Alloys Based on the Crystallographic Structure of Precipitates, *Furukawa Review,* Vol. 26, p. 39 -43.

Hall, E.O., (1951). *Proc. Phys. Soc. B,* Vol. 64, p. 747.

Heikkinen, S., (2010). Doctorate Thesis, Helsinki University.

Hofmann, U., Bögel, A., Hölzl, H., Kuhn, H.-A.: (2005) Beitrag zur Metallographie der Kupferwerkstoffe, *Praktische Metallographie, Vol.* 42, p. 329.

Höppel, H.W., Zhou, Z.W., Kautz, M., Mughrabi, H., Valiev, R.Z., (2002). In : Fatigue 2002 (Ed. A. Blom), EMAS, Stockholm, p. 1617.

Huchinson, B., Sundberg, R., Sundberg, M. (1990). *High Strength –High Conductivity Alloys,* Cu´90 Copper tomorrow (conf. proc.) Västeras, Oct 1 to 3, p. 245

Kinder, J., Huter, D., (2009). TEM-Untersuchungen an höherfesten elektrisch hochleitfähigen CuNi2Si-Legierungen, *Metall,* Vol. 63, p. 298.

Kuhn, H.-A., Nothhelfer-Richter, R., Futterknecht, G., (2000). Abhängigkeiten der dynamischen Elastizitäts- und Schubmoduln niedriglegierter Cu-Bandwerkstoffe, *Metall, Vol.* 54, p. 575-581.

Kuhn, H.-A., Käufler, A., Ringhand, D., Theobald, S. (2007). C7035 - a New High Performance Copper Based Alloy for Electro-Mechanical Connectors, *Materialwissenschaft und Werkstofftechnik,* Vol. 38, p. 624-636

Laird, C., Smith, G.C., (1962)., Crack propagation in high stress fatigue , *Philosophical Magazine,* Vol. 8, p. 847-857.

Larson, F.R., Miller, J., (1952). *Trans ASME,* Vol. 74, p. 765.

Lockyer, S.A., Noble, F.W. (1994). Preicipitate structure in a Cu-Ni-Si alloy, *Journal of Materials Science,* Vol. 29, p. 218.

Mandigo, F.N., Robinson, P.W., Tyler, D.E., Bögel, A., Kuhn, H.-A., Keppeler, F.M., Seeger, J. (2004) US Patent Application 0079456 A1.

Mutschler, R., Robinson, P.W., Tyler, D.E., Käufler, A., Kuhn, H.-A., Hofmann, U., (2009). WO Patent Application 2009/082695

Orowan, E., (1954). Dislocations and Mechanical Properties. Dislocations in Metals, (ed. by M. Cohen) (AIME: New York),p. 359- 377.

Qu, S., Zhang, P., Wu, S.D., Zang, Q.S., Zhang, Z.F., (2008). *Scripta Materialia,* Vol. 59, p.1131.

Randle, V., (1999). Mechanism of twinning-induced grain boendary engineering in low stacking fault energy material , *Acta Materialia,* Vol. 47, p. 4187.

Robinson, P.W. (2001). ASM Speciality handbook Copper and Copper Alloys (editor J.R. Davis), ASM International, Material Park, OH, p. 453.

Robinson, P., Gerfen, J., (2008). Copper Nickel Silicon Alloys for High Density Processor Sockets in: *Proc. Materials Science & Technology (MS&T),* Pittsburgh, Pennsylvania, p. 32.

Schuler, C.R., Kent, M.S., Deubner, D.C., Berakis, M.T., McCawley, M., Henneberger, P.K., Rossman, M.D., Kreiss, K., (2005). *American Journal of Industrial Medicine,* Vol. 47, p. 195.

Tanaka, K., Akiniwa, J., Kimura, H., (2002). In: Fatigue 2002 (Ed. A. Blom), EMAS, Stockholm, p. 1151.

Tyler, D.E. (1990). Wrought Copper and copper Alloy Products, Metal Handbook: 10th edit, Vol. 2 1990 (ASM) Ohio/USA- ISBN 0-87170-378-5, p. 245.

Vöhringer, O., (1972). Stapelfehlerenergie, Versetzungsdichte und- anordnung sowie Rekristallisations-Zwillingsdichte homogener Kupferwerkstoffe, *Metallwissenschaft und Technik,* Vol. 26, p. 1119.

Vöhringer, O., (1976). Verformungsverhalten von vielkristallinen α - Kupferlegierungen, *Metallwissenschaft und Technik,* Vol. 30, p. 1150.

Weidner, A., Amberger, D., Pyzcack, F., Schönbauer, B., Stanzl-Tschegg, S., Mughrabi, H., (2010). *International Journal of Fatigue,* Vol. 32, p. 872.

4

Lead-Free Wrought Copper Alloys for Bushings and Sliding Elements

Kai Weber and H.-A. Kuhn
Wieland-Werke AG, Ulm, Germany

1. Introduction

Components of copper and copper alloys are used in a wide range of applications in automobile manufacture and in machinery and plant construction. On the one hand, because of their high electrical conductivity, copper-based components are indispensable for the functionality of electrical and electronic units in motor vehicles and machines. On the other hand, they have a wide range of uses in the form of bushings and sliding elements. Bearings, bushings and sliding elements ensure transmission or conversion of the drive energy in machines, plants and in internal combustion engines. The installation situation of piston pin bore bushings in a combustion engine is pictured in Figure 1.

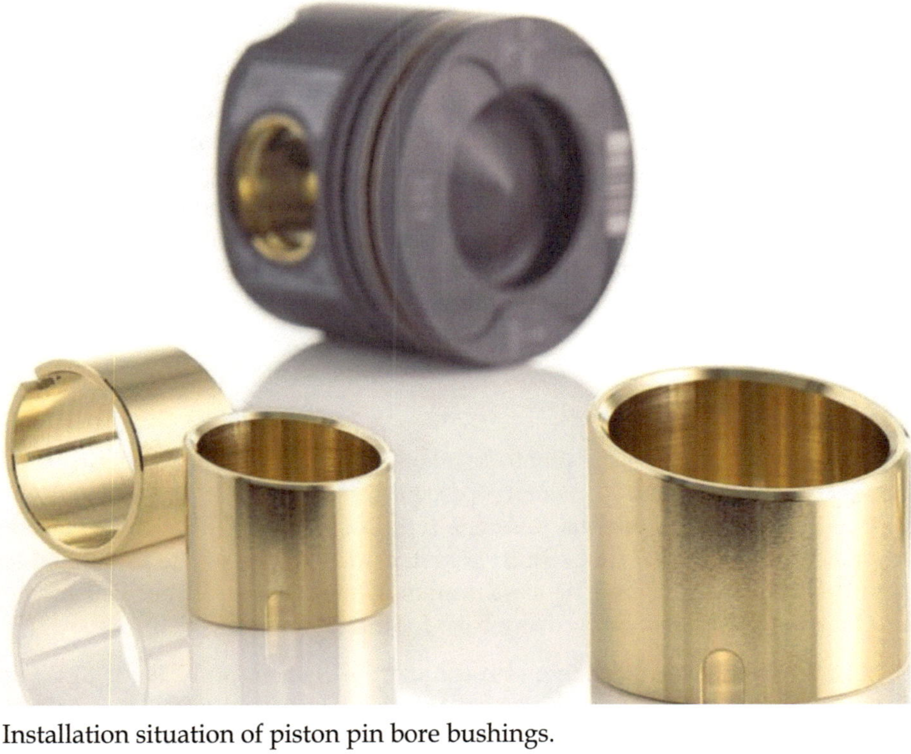

Fig. 1. Installation situation of piston pin bore bushings.

As a result of the increasing mechanical, tribological and thermal stresses to which the bearings in modern engines and plants are subjected, the materials used until now are increasingly reaching the limits of their stability under load. Wrought special brasses, multiphase tin-bronzes and nickel-tin-bronzes are the preferred metals.

Exactness of bushing dimensions made of wrapped strips or drawn tubes is realized by final machining. Beside the use of suited cutting tools and chose of appropriated machining parameters the productivity of chip removal depends also on the microstructure of the alloy. In customized special brasses lead forms a drop-like metallurgical phase which enables easy and economically machining of sliding elements like bushings in their final production stages. In use lead also improves lubrication behavior of sliding elements. The End-of-Life Vehicle Directive of the European Community intends to banish lead as an alloying element for metals. By minimizing the lead content in industrial products the legislator takes care in health of his residents. Manufacturer and supplier of bushings, bearings and sliding elements feel compelled to develop and offer lead-free materials.

Beside legislative demands on chemical compositions the development of new bushing materials is also driven by continuous improvements of engine performances. Bushings have to resist higher ignition points and cylinder pressures. Preferred copper-based wrought materials are special brasses, multiphase tin-bronzes and nickel-tin-bronzes. In addition lead-free substitutes have to meet requirements on economical machining. Due to a dramatic increase of copper prices the automotive industry asks for bushings made of less expensive metals. One technical solution to overcome materials costs is to replace sliding elements made of bulk materials by plated metal composites with only a thin wear resistant layer of copper alloy.

On the basis of a systematic analysis of the individual types of wear and damage mechanisms, this documentation will concentrate on the development of a new generation of lead-free brasses and tin-bronzes with intermetallics, copper-nickel-tin-alloys and roll clad composites made of spinoidally hardened bronzes. The development of wear resistant and thermally stable microstructures is discussed by assessing the roles of chemical elements forming metallurgical phases. Principles of process routes are explained. The resulting mechanical properties between room and elevated temperature and their thermal stabilities are presented by comparison with some older established brasses and bronzes. Wear resistance is evaluated by tribometer runs. Chip removal rates are compared to lead containing materials.

2. Wear and wear mechanisms

To allow the material of a wearing part to be adapted to the prevailing operating conditions, it is necessary to survey the elementary processes in the respective tribological system (Burwell, 1957/58). The numerous different types of stress to which the bushings and sliding elements are exposed first require a systematic analysis of the individual types of wear and damage mechanisms. The classification of the main wear mechanisms provided by Zum Gahr in Figure 2 will be used as a basis for this (Zum Gahr, 1988).

The lubricating film between a sliding element and the element on which it slides does not lead to a complete separation of the sliding surfaces in all operating states. The resultant mixed friction causes the bearings to be subjected to stress in the form of adhesion (fretting wear).

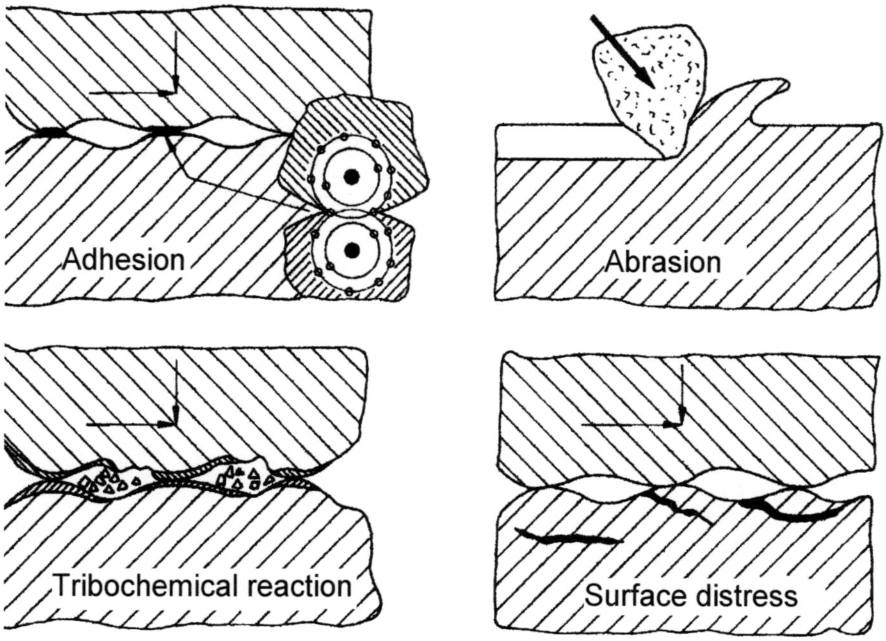

Fig. 2. Main wear mechanisms (Zum Gahr, 1988).

In the case of a plain bearing, this wear mechanism ultimately causes seizing of the shaft and consequently total failure (Bartel et al., 2004). A directionally oriented microstructure of the bearing material can decisively reduce its fretting tendency. Phases with body-centered cubic and hexagonal lattice structures are considered to be structural constituents that offer greater resistance to adhesion. On the other hand, structural elements with a face-centered cubic structure are more susceptible to fretting corrosion. The legally required abandonment of lead as an alloying element, responsible for improving the emergency running properties of the sliding elements, has focused attention on the wearing properties of the individual phase constituents.

Particles caused by dirt or abrasion can damage the running surface of the sliding element in various ways by abrasion (scoring wear or grooving) (Figure 3). To reduce the consequences of wear in the form of microcutting and microploughing, hardness and yield stress $R_{p0.2}$ of the bearing material are significant factors. In addition, the occurrence of the mechanisms of microfatigue and microcracking make its toughness properties important (Zum Gahr, 1992). Consequently, a heterogeneous microstructure with hard phases incorporated in a ductile matrix would be advantageous for the bearing material.

The frictional stress leads to the formation of particles and layers between the surfaces that come into contact (tribochemical reaction). When they reach a critical thickness, the reaction layers tend to undergo brittle spalling, with the consequent formation of wear particles (Collenberg, 1991). However, such reaction layers are highly important for the operation of sliding elements, since, as so-called tribofilms, they increase the resistance of the bearing material to adhesion (Bartel et al., 2004).

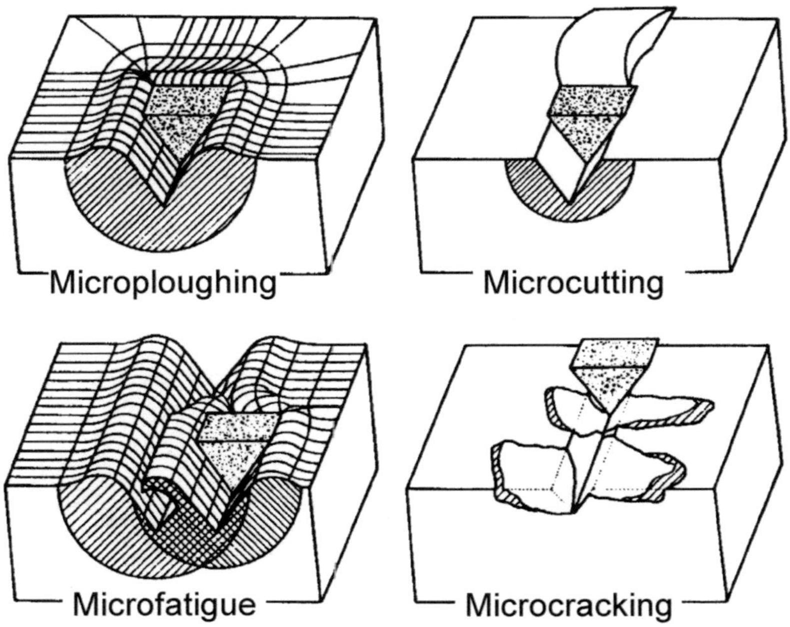

Fig. 3. Material damage under abrasive wear (Zum Gahr, 1992).

The varying hydrodynamic lubricating film pressure within the bearing points causes the sliding layer to undergo alternating tensile/compressive stress. Cracking that occurs leads to surface distress of the bearing (fatigue wear). The material should therefore have a heterogeneous microstructure with a ductile matrix in order to have a high resistance to crack initiation and propagation.

In addition to the wear mechanisms, the sliding elements in internal combustion engines and in many machines and plants are exposed to thermal stress. This may lead to undesired changing of the microstructure and to stress relaxation of the component. The accompanying loss of strength and dimensional stability limits the operational reliability of the sliding elements. To increase the thermal stability of the microstructure, the proportion of phases with greater transformation activity should be limited. Furthermore, fine precipitates should reduce the extent of thermally induced stress relaxation of the sliding element.

3. Experimental procedures

3.1 Characterization of microstructure

Grains, grain boundaries and metallurgical phases were examined by optical light microscopy. Specimen preparation by immersion etching in a sulphuric acid solution of $K_2Cl_2O_7$ was first described by Schrader (Schrader, 1941). The complete method was also reported in detail by Kuhn et al. (Kuhn et al., 2004) and Hofmann et al. (Hofmann et al., 2005).

3.2 Mechanical properties

Yield strength, ultimate tensile strength and ductility A5 were determined at room temperature in accordance with standard EN 10002 by Z100 (Zwick company). Mechanical properties at elevated temperatures up to 400°C were examined by a Zmart.Pro tension test machine (ZwickRoell company) equipped with an irradiation furnace (Maytec company). Hardnesses at room temperature were measured by Brinell Hardness HB.

3.3 Tribological properties

When choosing a test method, the stress conditions actually occurring should be approximated (Römer & Bartz, 1981; Czichos & Habig, 2003; Grün et al., 2007). Therefore, the ring segment/disk modeling method was used for examining the adhesive wear behavior of the bushing materials. These materials were represented by the ring segments. A disk of 100Cr6 was operating as the counterpart of this setup. Implementing this type of method as a fretting test allows the loading limit of the sliding material to be determined by means of increasing the load in intervals. The ring segment/disk arrangement was chosen to initiate conditions in real bearing assemblies as described by Pucher et. al. (Pucher et al., 2009). The complete test conditions used are listed in Table 1.

Tribometer	Wazau TRM 1000
Disk material	100Cr6
Ring material	Bushing materials
Lubrication	Bath lubrication with engine oil Shell Rimula SAE30
Starting temperature	120°C
Running-in phase	5 min at 200 N
Load increase	by 50 N every 2 min
Maximum load	Tubes: 1050 N Strips: 1200 N
Maximum surface pressing	16 N/mm^2
Sliding velocity	1 m/s
Cutout criteria	Torque > 2,5 Nm or maximum load is reached
Measured variables	Friction moment, temperature, wear

Table 1. Test conditions of the ring segment/disk arrangement.

4. Alloys for wear-resistant applications

4.1 Brass materials

The two high-strength brasses CuZn31Si1 (Wieland designation SB8) and CuZn37Mn3Al2PbSi (Wieland designation S40) were taken as the starting point for the development of a new wear-resistant copper alloy.

The microstructure of CuZn31Si1 has an adequate stability up to temperatures of about 200-250°C. On the other hand, the adhesive wear resistance of components made of CuZn37Mn3Al2PbSi is excellent.

For the development of a new brass-based copper alloy for sliding elements, the objective was therefore the combination of the adequate thermal stability properties of Wieland-SB8 and the very good sliding properties of Wieland-S40.

The above analysis of the effective wear mechanisms and the possibilities of using aspects of materials technology to influence the extent of wear damage formed a fundamental basis for the conceptual development of the new bearing alloy CuZn31Mn2Al1Ni1Si1 (Wieland designation SX1 (Weber, 2009).

In the following section a comparison of the alloy composition, microstructure and properties of these three bearing materials (Weber & Pucher, 2009) is presented.

4.1.1 Chemical composition and microstructure

Table 2 lists approximate values for the chemical composition of the two reference materials Wieland-SB8 and Wieland-S40 as well as the alloy content of the new high-strength brass Wieland-SX1.

Alloy	Cu	Zn	Pb	Si	Mn	Ni	Fe	Al
Wieland-SB8	68	Remainder	< 0.3	1	-	-	-	-
Wieland-S40	58	Remainder	0.5	1	2	0.5	0.5	2
Wieland-SX1	64	Remainder	-	1	2	1	0.5	1

Table 2. Composition of the bearing materials (approximate values in % by weight).

On account of the high Cu content and the absence of Mn, Wieland-SB8 is equivalent to an α-β brass (Figure 4). In the α-matrix, the dark-colored β phase components in the form of rows can be found.

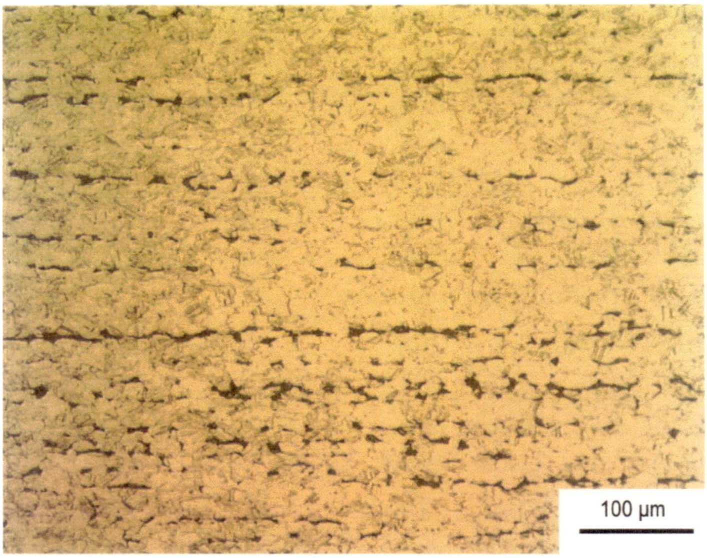

Fig. 4. Microstructure of Wieland-SB8 (magnified 200 times).

It is clear from the micrograph of Wieland-S40 that, owing to the comparatively low Cu content, the matrix is made up almost exclusively of the β phase (Figure 5). This is interspersed with silicides of the modification Mn_5Si_3, shown in gray.

The very fine microstructure with the network of β phase islands (yellow) incorporated in the α matrix and with Mn mixed silicides of various sizes and forms is characteristic of the new bearing material Wieland-SX1 (Figure 6). This microstructure combines the feature of a lower β content of the alloy Wieland-SB8 with the content of hard silicides of the Wieland-S40. Greater magnification additionally reveals very fine silicides that are primarily embedded in the α matrix (Figure 7).

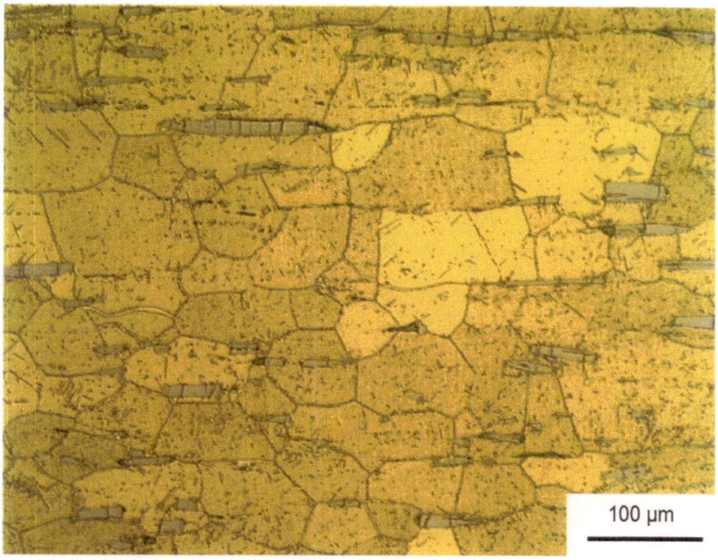

Fig. 5. Microstructure of Wieland-S40 (magnified 200 times).

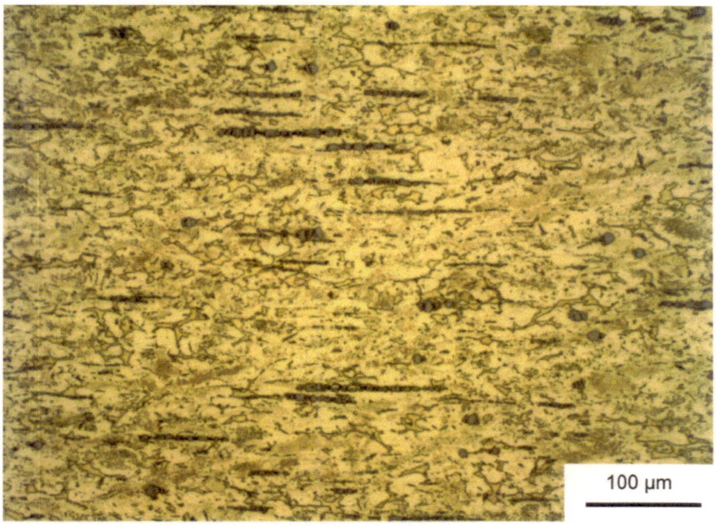

Fig. 6. Microstructure of Wieland-SX1 (magnified 200 times).

4.1.2 Mechanical properties

Among the forms of raw material used for machining the sliding elements are bars or tubes that are produced by means of the casting/hot forming/cold forming + annealing processes. The mean values of the most important mechanical properties of tubes from a variety of batches are represented in Table 3.

Alloy	Wieland-SB8	Wieland-S40	Wieland-SX1
Raw material	Tube 30.1×24.7 mm	Tube 60×48 mm	Tube 30.1×24.7 mm
HB	195	163	207
R_m [MPa]	654	628	648
$R_{p0.2}$ [MPa]	593	315	566
A5 [%]	14.4	21.4	13.7

Table 3. Mean values of some of the mechanical properties of the bearing materials.

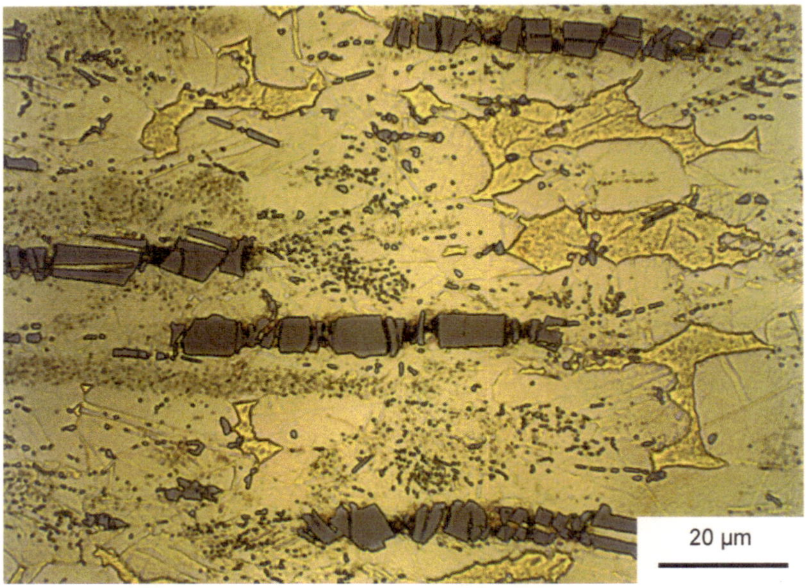

Fig. 7. Microstructure of Wieland-SX1 (magnified 1000 times).

In particular, the value for the hardness HB and the yield point $R_{p0.2}$, which are so important for sliding elements, of the Wieland-S40 tubes are at a comparatively low level. With regard to these two properties, Wieland-SX1 ensures a close approximation to the maximum values of the comparative materials. This allows the aim of setting adequate mechanical characteristic values to be regarded as achieved.

The high hardness represents the basic prerequisite for an appropriate resistance to the abrasive damage mechanisms of microploughing and microcutting (Figure 3). Furthermore, the microstructure of Wieland-SX1 is characterized by a ductile matrix with Incorporated silicide phases. This heterogeneous structure guarantees a high resistance to microfatigue and microcracking as well as to surface distress.

4.1.3 High-temperature strength

For the bearing materials under consideration, Figures 8 and 9 show the change in the tensile strength R_m and the yield strength $R_{p0.2}$ in the course of an increase in temperature up to 400°C.

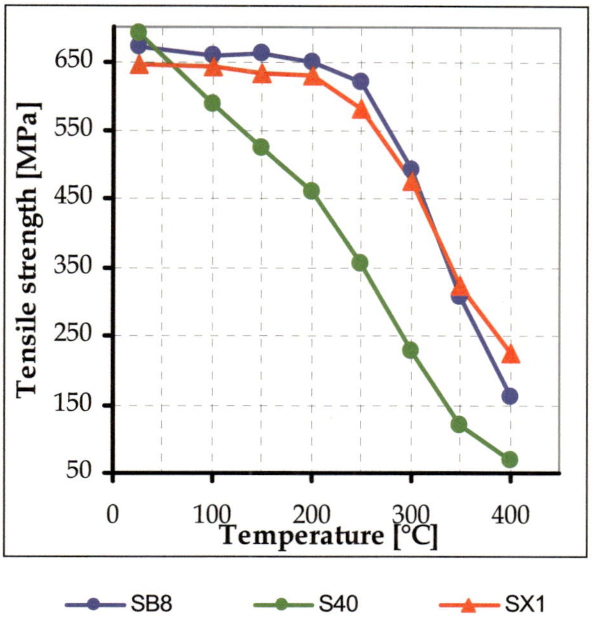

Fig. 8. Tensile strength R_m as a function of temperature.

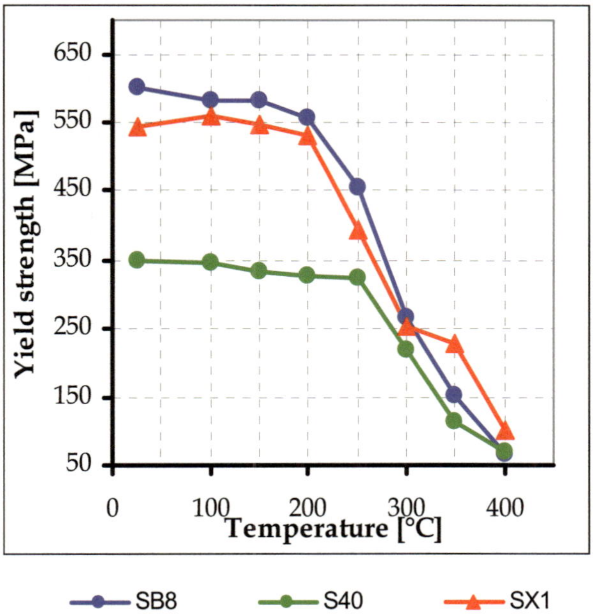

Fig. 9. Yield strength $R_{p0.2}$ as a function of temperature.

As the temperature rises, a clear drop in R_m and $R_{p0.2}$ of the β-rich alloy Wieland-S40 can be seen. This is caused by the thermally induced transformation of the β phase into the α phase. As a result of the α-poorer microstructure of the bearing materials Wieland-SB8 and Wieland-SX1, the reduction in the strength values in the case of these alloys during a temperature increase up to about 200-250°C is smaller. It is therefore evident that the new alloy Wieland-SX1 meets the requirements for the thermal stability of the microstructure.

4.1.4 Wear properties

Figure 10 shows the variation in the friction coefficient as a function of the load and running time of the test for the three bearing materials. Owing to the predominant component in the microstructure represented by the α phase with a face-centered cubic structure in the Wieland-SB8 alloy, even a small load causes fretting of the parts co-acting in the bearing. On the other hand, the Wieland-S40 matrix comprising the β phase (body-centered cubic structure) together with the Mn_5Si_3 silicides of a hexagonal structure ensures the comparatively highest resistance to adhesive wear.

The lowered β content in the microstructure of Wieland-SX1 in comparison with Wieland-S40 leads to a greater running-in phase. However, the high proportion of larger Mn_5Si_3 silicides and the tribological stabilization of the α phase provided by the finer silicides prevent fretting of these specimens even under maximum load. This slight lowering of the friction coefficient from about halfway through the test is an indication of the greater importance of the formation of a tribofilm for the adhesive wear resistance in comparison with Wieland-S40. This separating layer between the metallic contact surfaces together with the advantageous heterogeneous microstructure of the alloy provides the prospect of a stable running behavior of the Wieland-SX1 sliding elements even as the test continues.

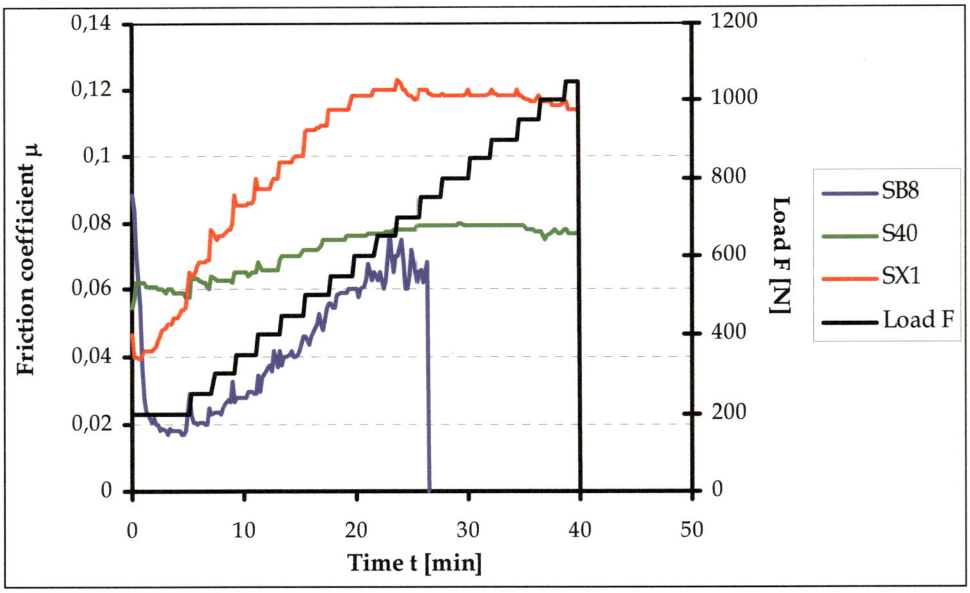

Fig. 10. Variation in the friction coefficient as a function of load and running time.

It is therefore possible to state that, even without Pb, the new Wieland-SX1 bearing alloy has the necessary resistance to fretting corrosion.

4.1.5 Summary

The conventional plain bearing materials in internal combustion engines are increasingly reaching their performance limits. The tribological and thermal stresses are increasing because the demands for lightweight construction are causing the dimensions of the components to be reduced and at the same time the requirements for lower emissions are causing the ignition pressures to be increased. Furthermore, the use of low-viscosity oils is leading to an increase in states of wear-intensive mixed friction, which in the case of sliding elements in the engine compartment occur especially during starting and stopping operations and when the lubricating film breaks down.

The alloying systems CuZn31Si1 and CuZn37Mn3Al2PbSi have either high-temperature strength or extremely great resistance to fretting corrosion. Furthermore, both bearing materials contain lead as an alloying constituent. According to the EU End-of-Life Vehicle Directive, however, this alloying element, which is considered to be toxic, will be banned in the future.

Therefore, the objective was to develop an alternative, Pb-free brass material which at the same time meets the requirements for thermal and tribological material properties.

The heterogeneous form of the microstructure with the network of β phase islands incorporated in the ductile α matrix and with Mn-silicides and mixed silicides of various sizes and forms lends the new bearing alloy Wieland-SX1 a high degree of thermal stability and complex wear resistance. By also conforming to environmental guidelines, this material is especially suitable for being used as a sliding element in future internal combustion engines, transmissions, ancillary units, drive trains and brakes.

4.2 Tin-bronzes

The clean metal and alloy surfaces in contact exhibit high adhesion, and consequently high friction and wear. The coefficient of friction of contacting metallic surfaces cleaned in a high vacuum can be very high, typically 2 and much higher (Bhushan, 2002).

Most metals and alloys oxidize in air to some extent and form chemical films across the interface. In the tin-bronzes, the alloying element Tin form oxides very rapidly. The tin-oxide film acts as a low shear-strength film and in addition because of low ductility leads to low friction. The oxide film may effectively separate the two metallic surfaces. Furthermore, the oxides increase the resistance against corrosion.

However, during sliding at higher loads, the thin oxide film may be penetrated, can come off and transition occurs to high values of friction and wear. For this reason, the positive wear resistance of the tin-bronzes can be further improved by alloying the bronzes with other elements. Consequentially new generations of wear resistant tin bronze were developed.

The following sections describe the microstructure, the mechanical and the adhesive wear properties of three classes of Cu-Sn-alloys that are listed in Figure 11.

Cu-Sn-bronze	Cu-Ni-Sn-bronze	Cu-Sn-bronze with silicide phases HSn
CuSn8 (Wieland-B09)	CuNi6Sn6 (Wieland-L66®)	(temporary name while under development)

Fig. 11. Classes of Cu-Sn-alloys for friction bearing applications.

4.2.1 Chemical composition and microstructure

The wrought Cu-Sn-P-alloy CuSn8 (Tin-Bronze, Phosphor-Bronze) is characterized by an α-solid solution. The chemical composition is shown in the table 4. Fig. 12 describes the microstructure of this material (Wieland designation B09).

A representative of the Cu-Ni-Sn-alloy series is the bronze CuNi6Sn6 (Wieland designation L66®). Wieland-L66® is a spinoidally hardened bronze with the composition shown in table 4. Figure 13 illustrates the fine-grained microstructure with discontinuous Ni-Sn-precipitations, which are located predominantly at the grain boundaries.

Alloy	Cu	Sn	P	Mn	Ni	Fe	Si	Zn	Al
Wieland-B09	Remainder	8,0	0,2	-	-	-	-	-	-
Wieland-L66®	Remainder	5,5	-	-	6,0	-	-	-	-
HSn	Remainder	4,3	-	1,3	-	0,8	0,6	2,5	0,8

Table 4. Composition of the lead-free types of Cu-Sn-alloys (approximate values in % by weight).

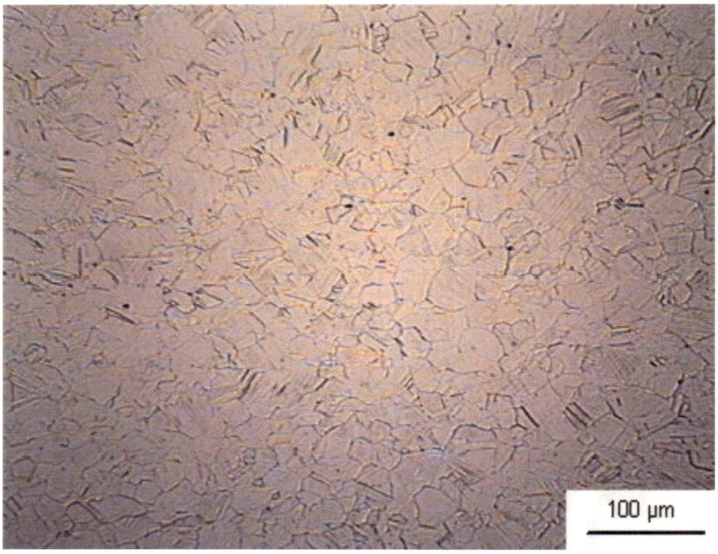

Fig. 12. Microstructure of Wieland-B09 (magnified 200 times).

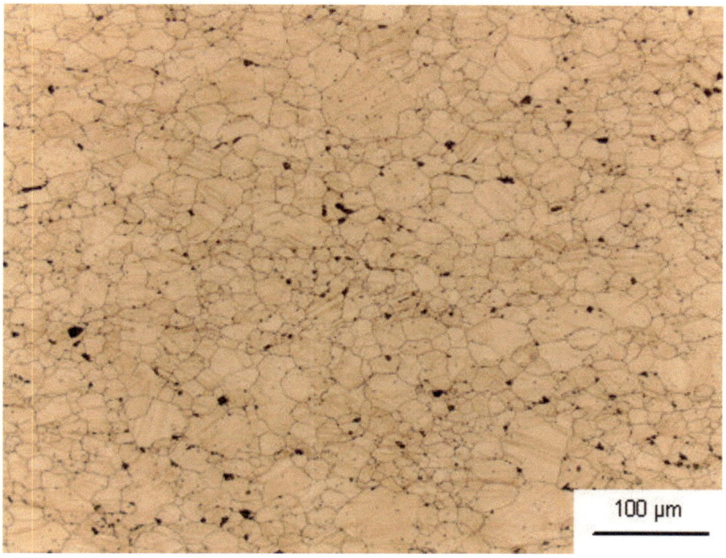

Fig. 13. Microstructure of Wieland-L66® (magnified 200 times).

The silicide phases in the special brass Wieland-SX1 ensures higher strength and wear resistance. For this reason, a new class of tin bronzes with silicides (Wieland temporary designation HSn) has been developed. The composition of a representative of this class is shown in table 4. Figures 14 and 15 illustrate the finest microstructure with mixed silicides in different sizes and shapes.

Fig. 14. Microstructure of HSn (magnified 200 times).

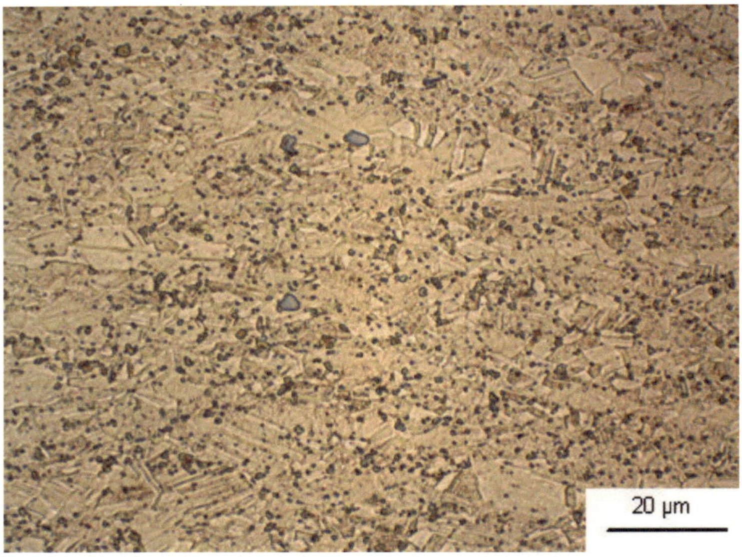

Fig. 15. Microstructure of HSn (magnified 1000 times).

4.2.2 Mechanical properties

As the solid solution hardening effect of tin in copper is strong, high strength and hardness values of the tin-bronze Wieland-B09 can be achieved without the need for precipitation hardening (table 5). The spinoidally hardened nickel-tin-bronze Wieland-L66® show superior strength and hardness. In the HSn-bronze, the silicide phases in a ductile matrix lead to a combination of higher strength, hardness and best toughness. Useful for comparison, the mechanical properties of the Wieland-SX1 which is combined to form a strip as shown in table 5.

Alloy	Wieland-B09	Wieland-L66®	HSn	Wieland-SX1
Raw material	rod (hard temper)	strip (hard temper)	strip (hard temper)	strip (hard temper)
HB	150	245	210	200
R_m [MPa]	570	780	650	611
$R_{p0.2}$ [MPa]	420	700	540	500
A5 [%]	23	12	28	13

Table 5. Mean values of some of the mechanical properties of the tin-bronzes.

4.2.3 High-temperature properties

For the tin-bronzes Wieland-B09, Wieland-L66® and HSn, Figures 16 and 17 show the change in the tensile strength R_m and the yield strength $R_{p0,2}$ in the course of an increase in temperature up to 400°C.

Precipitation hardened nickel-tin-bronze Wieland-L66® show superior thermal stability in comparison with conventional tin-bronze Wieland-B09 and with silicides-containing bronze HSn.

Mounted bushings tend to creep due to mechanical stresses and thermally induced stresses. Depending on the piston diameter loads on bushings achieve 100 to 300 kN at temperature up to 200 °C. In terms of creep these load can be defined by a range of 10^{-3} to 10^{-4} of shear stress normalized to shear modulus and simultaneously by homologous temperatures of 0,25 to 0,4. Cu-Ni-Sn-alloys with 5 and more wt% of each alloying element Ni and Sn exhibit excellent resistance against stress relaxation. For example, the stress relaxation rate of CuNi6Sn6 at 200 °C is only 0,2 MPa/h compared with 1 MPa/h for the brass CuZn31Si1 (Kuhn et al., 2007).

Figure 18 characterizes stress relaxation of the two finally relief annealed bushing materials CuSn8 and CuNi6Sn6 between 1 and 3000 hours under load at 200 °C. The loss in stress $\Delta\sigma$ is expressed in percentage of the initially applied stress. For this experiment the initial stresses were chosen in the order of 50% of the yield stresses. The long term relaxation resistance was extrapolated from short term experiments which were performed on strips via the ring method (Bögel, 1994). The extrapolations of losses in initial stresses $\Delta\sigma_{rel}$ were calculated by means of Larson-Miller parameter P (Larson, 1952).

$$\log\Delta\sigma_{rel} = a\,P - b \qquad (1)$$

(a, b : coefficients of the regression curves)

$$P := (T/°C + 273.16)\,(20 + \log_{10}(t/h))\,10^{-3} \qquad (2)$$

For the precipitation hardened CuNi6Sn6 shows 80% remaining stress after 3000 hours whereas bushings made of solid solution hardened CuSn8 will totally fail after this time.

This superior resistance of CuNi6Sn6 against stress relaxation is caused by spinodal decomposition of solid solution annealed and subsequently cold rolled strips of CuNiSn-alloys in a temperature range of 200 to 400 °C (Plewes, 1975).

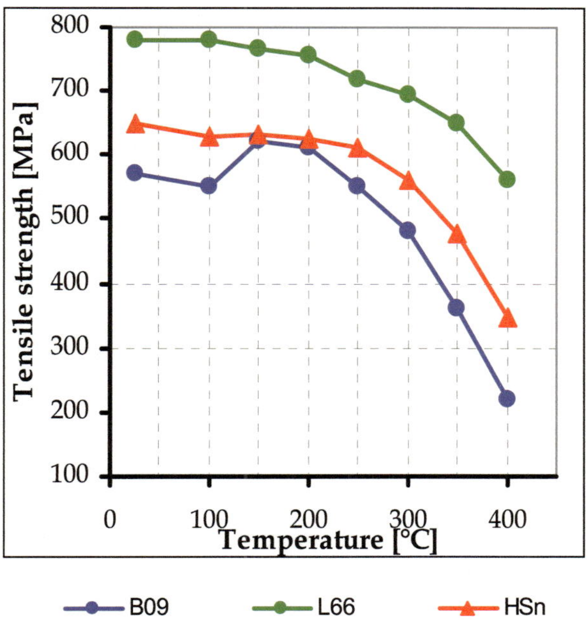

Fig. 16. Tensile strength R_m as a function of temperature.

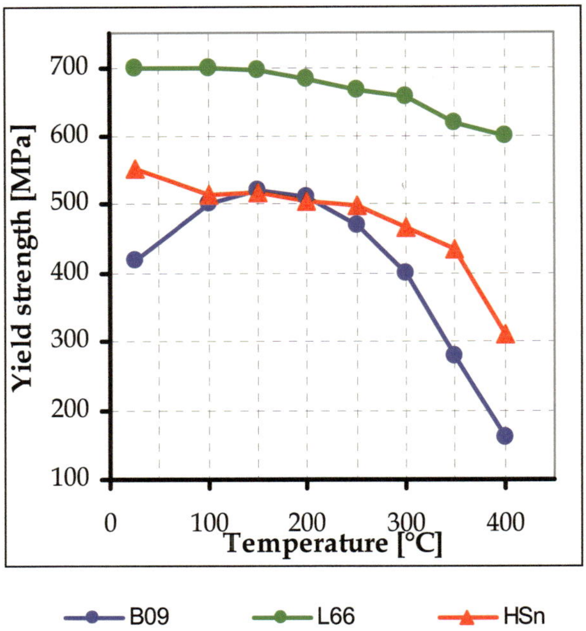

Fig. 17. Yield strength $R_{p0.2}$ as a function of temperature.

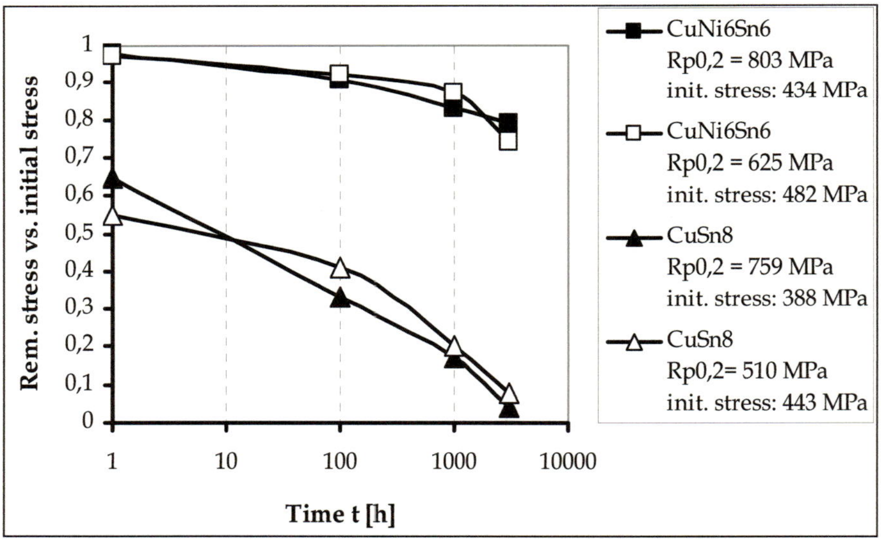

Fig. 18. CuNi6Sn6 and CuSn8: long term forecast of resistance against stress relaxation at 200°C.

4.2.4 Wear properties

The wear resistance of strips of the three bronze types Wieland-B09, Wieland-L66® and HSn is evaluated by tribometer runs. The measured variables include the friction coefficient μ and the wear rate w that constitutes the thickness loss of the strips during the measurement time. For mapping the scatter of the measured results, figures 19-22 shows the especially revealing variation in the friction coefficient and in the wear rate of two specimens 1 and 2.

Figure 19 shows the variation in the friction coefficient and in the wear rate of two specimens of the bronze CuSn8. Remarkable is a significant increase in the thickness loss of the strip.

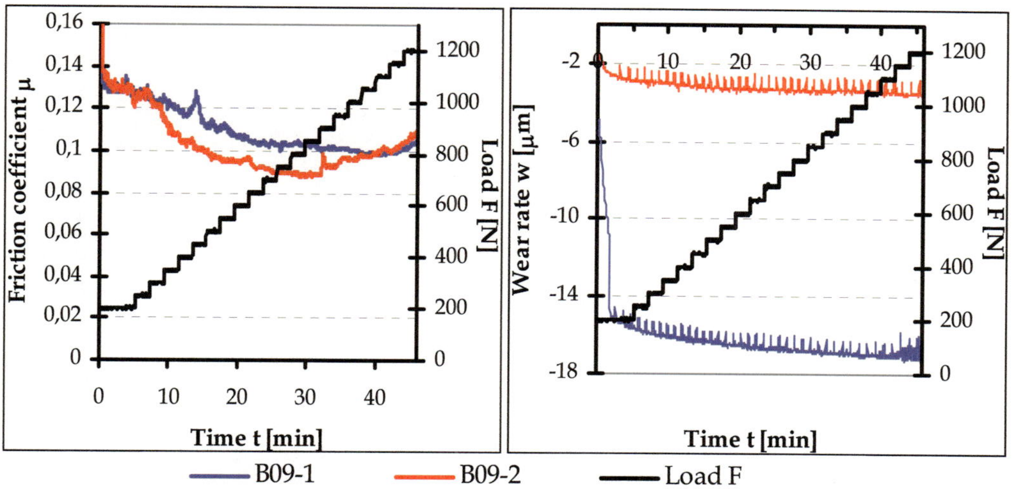

Fig. 19. Variation in the friction coefficient and in the wear rate of two strip-specimens of the Cu-Sn-bronze Wieland-B09 as a function of load and running time.

The bumpy running behavior of the CuNi6Sn6-specimens can be found in figure 20. During the longer start-up phase, the friction coefficient decreases. However, the wear rate is further increased.

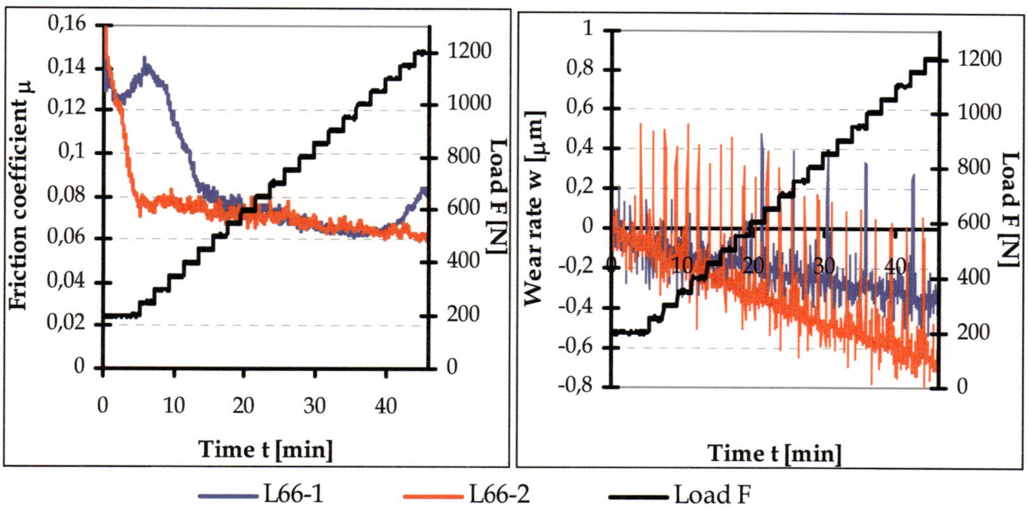

Fig. 20. Variation in the friction coefficient and in the wear rate of two strip-specimens of the Cu-Ni-Sn-bronze Wieland-L66® as a function of load and running time.

In comparison to the CuNi6Sn6, the specimens of the bronze HSn guarantee a smooth running (Figure 21). The wear rate is extremely low. The more ductile matrix of the HSn stands more dirt and abraded particles than CuNi6Sn6. That is why the wear rate can be positive (see HSn-1).

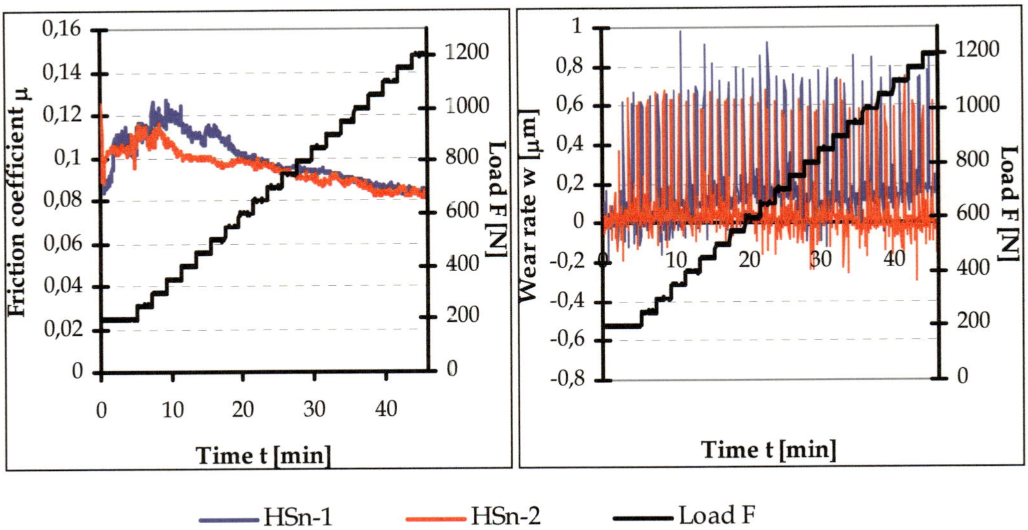

Fig. 21. Variation in the friction coefficient and in the wear rate of two strip-specimens of the new bronze HSn as a function of load and running time.

The smooth running behavior, the good friction coefficient and the very low wear rate of the HSn-bronze are taking place as a result of the heterogeneous microstructure with hard silicides in a ductile matrix. The silicide content of the alloy results in good adhesive wear resistance. The matrix with higher strength and excellent toughness means that the silicides cannot be excavated and broke out from the surface.

These interconnections are also evident in the results of the wear test of the Wieland-SX1-specimens. The silicides with different size and shape in the mixed α-β-matrix guarantee a good wear resistance (figure 22).

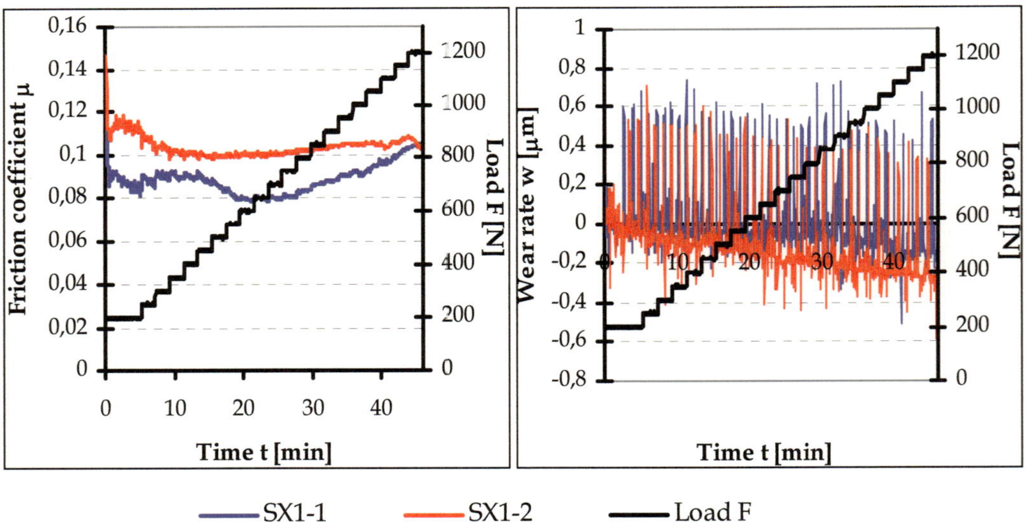

Fig. 22. Variation in the friction coefficient and in the wear rate of two strip-specimens of the new brass material Wieland-SX1 as a function of load and running time.

4.2.5 Summary

In addition to brass alloys, tin-bronzes can be used for bushings, bearings and sliding elements. The Cu-Sn-alloys show a large number of positive material properties. These alloys have a high strength and a good resistance against wear and corrosion. However, during sliding at higher loads, the resistance to wear is inadequate for most sliding applications. For this reason, the good wear resistance of the tin-bronzes can be further improved by alloying the bronzes with other elements. The result of this action is the development of new generations of tin-bronzes - the Cu-Ni-Sn-bronzes and the Cu-Sn-alloys with silicide phases.

The spinoidally hardened Ni-Sn-bronze Wieland-L66®, which is outstanding for its particularly superior strength at room and at elevated temperatures, is characterized by a low level of friction coefficient for the duration of the wear test. However, the results show that the highest strength values are not decisive alone. The wear rate of the CuNi6Sn6-specimens increases over running time. The reason for that is the too low toughness of this material. The wear mechanisms surface distress and micro cracking cause an increased wear.

The HSn, a newly developed tin-bronze with silicides, offers a unique combination of high strength, hardness, toughness and wear resistance. The silicide phases give the material very good sliding properties. Furthermore, the matrix with good hardness and very high ductility means that the silicides cannot be excavated and broke out from the surface. This heterogeneous microstructure ensures an excellent resistance against abrasive and adhesive wear. Also noteworthy is the compatibility for the dirt and abraded particles. The result is that the friction coefficient will not improve during operation.

The HSn and Wieland-L66® are particularly suitable for sliding elements in prospective internal combustion engines, transmissions, ancillary units, drive trains and brakes.

5. Production process of monometallic bushings

Wrapped bushings are produced from the following steps: continuous casting of slabs, hot rolling, cold rolling, cutting according to width, punching, stamping, rolling and calibrating (Fig. 23). Machined bushings result from hot extrusion of tubes, drawing, turning and grinding.

6. Production process and properties of friction bearings in steel-composite construction

Since a few years the metal prices of copper, tin and nickel have dramatically increased. Manufacturer of bushings and their customers look for appropriate replacement of monometallic sliding elements. Without renouncing the excellent combination of strength, stress relaxation resistance and wear behavior one accepted technical solution are roll clad metal composites for connecting rods. Via cold rolling (Figure 24) and diffusion annealing a thin strip of CuNi6Sn6 are plated onto a ferritic steel strip (Ababneh et al., 2006).

Adhesion of both materials can be improved by soft interlayer of pure copper. Figure 25 describes a composite of a steel strip, an intermediate Copper layer and a CuN6Sn6 strip. For a successful roll cladding a true strain φ of -0.7 of each material is needed.

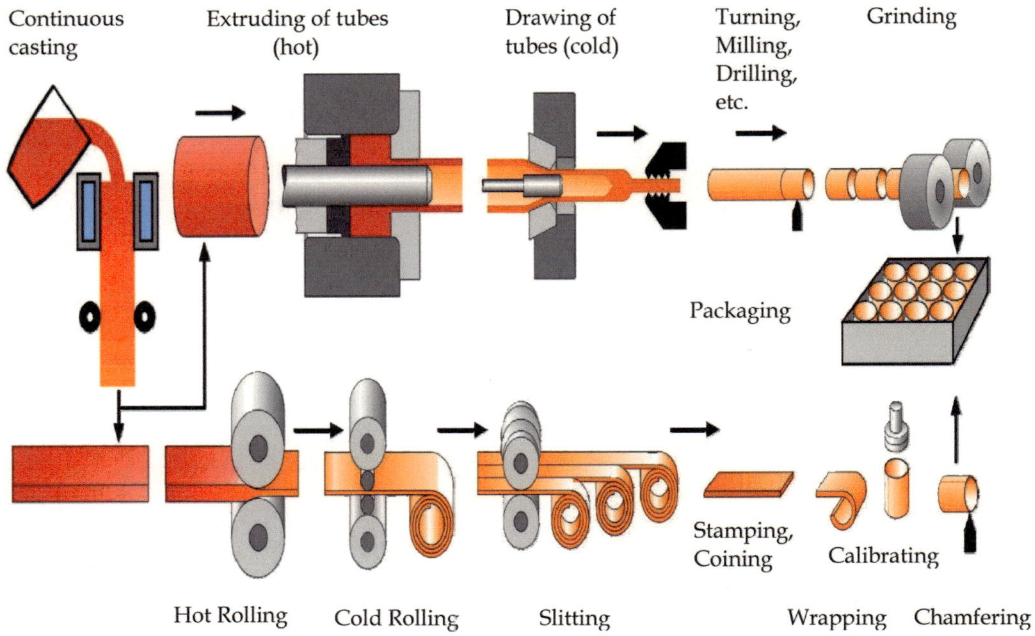

Fig. 23. The manufacturing process from casting to the finished machined and wrapped bushing (Scharf, 1999).

Fig. 24. Principle of roll cladding of a three layer metal composite.

The mechanical strength of the clad metal composite is given by the steel layer and the CuNi6Sn6 in the age hardened stage. In addition the CuNi6Sn6 layer is responsible for a good resistance against wear.

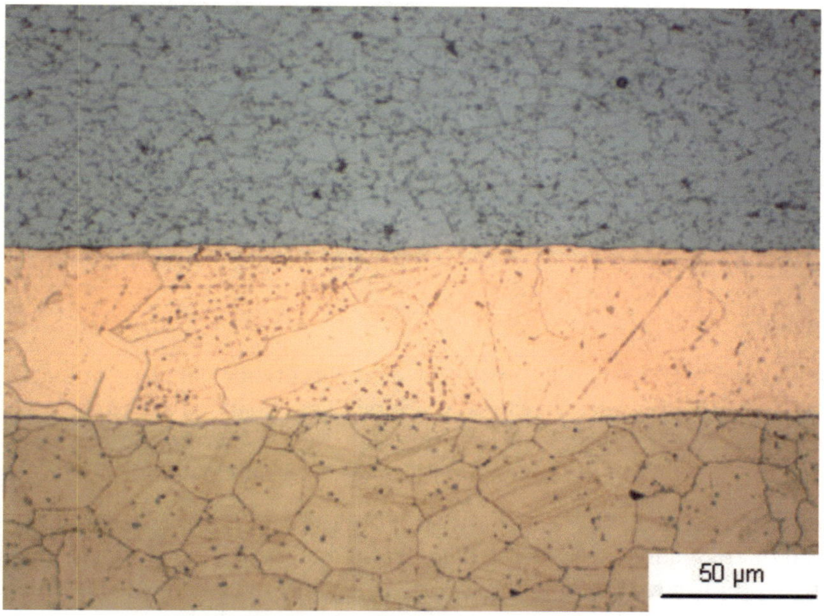

Fig. 25. Micrograph (longitudinal section) of roll clad wrapped bushing for connecting rods: ferritic steel (top), copper (middle), CuNi6Sn6 (bottom).

Figure 26 exhibit the response of a cold rolled clad strip on age hardening between 200°C and 460°C. The time of annealing was 3 hours. One-third of the composite is the nickel tin bronze and two-third is a 0,15 wt% ferritic steel. Annealing has no influence on hardness of the steel whereas the hardness of the age hardened CuNi6Sn6 layer increases in the order of 100% from 135 HBW1/30 to 270 at 380°C. Simultaneously, the strengths of the compound were improved by 30% of the room temperature values.

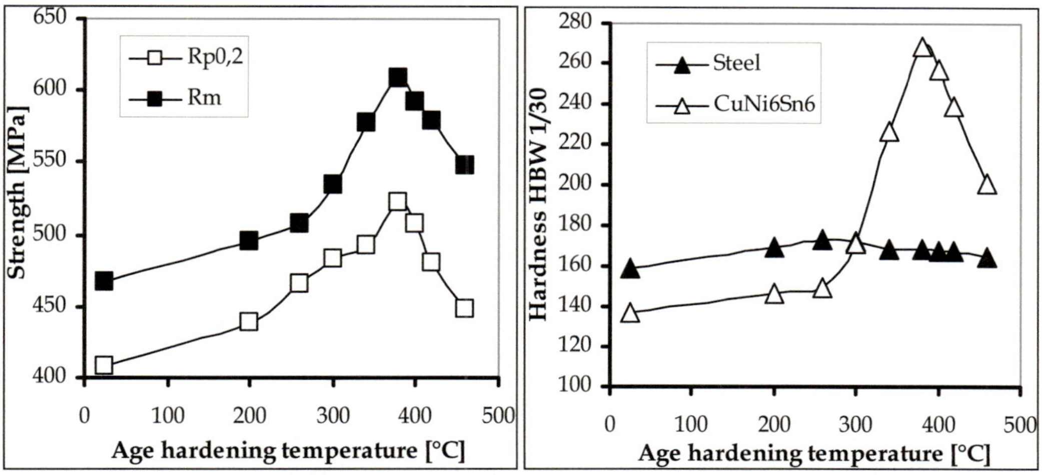

Fig. 26. Effect of age hardening on mechanical properties of a clad compound (left) and on hardness of single layer (right).

The spring back behavior of wrapped bushing designed by a clad compound is subjected to the Yield strength and the Young´s modulus E. The Young´s modulus of the composite was determined via tension test. In good accordance with a prediction by a rule of mixture (Reuss approach) with 2/3 of 210 GPa (=E_{steel}) and 1/3 of 125 GPa (=$E_{CuNi6Sn6}$) an $E_{Composite}$ of 175 MPa was measured. For this estimation of E the contribution of the thin intermediate copper layer was neglected.

Rule of mixture, constant strain approach (Reuss approach):

$$E_{Composite} = (3\ E_{steel}\ E_{CuNi6Sn6}) \ / \ (2E_{CuNi6Sn6} + E_{steel}) \quad (3)$$

7. Conclusion

Because of its good mechanical properties, its compatibility for the abrasion particles and its excellent behaviour under mixed friction conditions, copper-based alloys have a wide range of uses in the form of bushings and sliding elements. Preferred copper-based wrought materials are special brasses, tin-bronzes, multiphase tin-bronzes and nickel-tin-bronzes.

As a result of the increasing mechanical, tribological and thermal stresses to which the bearings in modern engines are subjected, the brass materials and the tin-bronzes used until now are increasingly reaching the limits of their stability under load. The tribological and thermal stresses are increasing because the demands for lightweight construction are causing the dimensions of the components to be reduced and at the same time the requirements for lower emissions are causing the ignition pressures to be increased. Furthermore, the use of low-viscosity oils is leading to an increase in states of wear-intensive mixed friction, which in the case of sliding elements in the engine compartment occur especially during starting and stopping operations and when the lubrication film breaks down. In addition, the End-of-Life Vehicle Directive of the European Community intends to banish lead as an alloying element for metals. This necessity has given rise to the development of novel Pb-free bearing alloys with the required combination of properties.

To allow the material of a wearing part to be adapted to the prevailing operating conditions, it is necessary to survey the elementary processes in the respective tribological system. The analysis of the effective wear mechanisms and the possibilities of using aspects of materials technology to influence the extent of wear damage formed a fundamental basis for the conceptual development of the new bearing alloys.

The two brasses CuZn31Si1 (Wieland designation SB8) and CuZn37Mn3Al2PbSi (Wieland designation S40) were taken as the starting point for the development of the new wear-resistant brass alloy Wieland-SX1. The heterogeneous form of the microstructure with the network of β phase islands incorporated in the ductile α matrix and with Mn-silicides and mixed silicides of various sizes and forms lends the new bearing alloy Wieland-SX1 a high degree of thermal stability and complex wear resistance.

In addition to the brass alloys, the tin-bronzes are used as material for bushings and sliding elements. In the tin-bronzes, the alloying element tin form oxides very rapidly. The tin-oxide film acts as a low shear-strength film and in addition because of low ductility leads to low friction. The oxide film may effectively separate the two metallic surfaces. Furthermore, the oxides improve the resistance against corrosion. However, during sliding at higher loads, the thin oxide film may be penetrated, can come off and transition occurs to high values of friction and wear. For this reason, the positive wear resistance of the tin-bronzes can be further improved by alloying the bronzes with other elements. Cu-Ni-Sn-alloys and the Cu-Sn-bronze with silicide phases were the results of these developments.

The spinoidally hardened alloy CuNi6Sn6 (Wieland designation L66®) shows superior mechanical properties, a good wear resistance and a superior resistance against stress relaxation. In the silicides-containing bronze (Wieland designation HSn), the silicide phases in a ductile matrix lead to a combination of higher strength, hardness and best toughness. The combination of these properties is the reason for the very good resistance against the wear mechanisms adhesion, abrasion and surface distress.

By also conforming to environmental guidelines, the alloys Wieland-SX1, Wieland-L66® and HSn are particularly suitable for being used as a sliding element in future internal combustion engines, transmissions, ancillary units, drive trains and brakes.

8. References

Ababneh, M.; Kuhn, H.-A. & Voggeser, V. (2006). Material Strip Form and it use, composite sliding element. *Patent application DE 102006019826.3*

Bartel, D.; Bobach, L. & Deters, L. (2004). Fresskriterien für ölgeschmierte Radialgleitlager. *Tribologie und Schmierungstechnik*, Vol. 51, No. 5, pp. 29-38

Bhushan, B. (2002). *Introduction to tribology*. John Wiley & Sons, New York

Bögel, A. (1994). Spannungsrelaxation in Kupferlegierungen für Steckverbinderwerkstoffe und Federwerkstoffe. *Metall*, Vol. 48, pp. 872-875

Burwell Jr., U.T. (1957/58). Survey of possible wear mechanism. *Wear*, Vol. 1, pp. 19-141

Collenberg, H.F. (1991). *Untersuchungen zur Fresstragfähigkeit schnelllaufender Stirnradgetriebe*. Dissertation, Technische Universität München, Germany

Czichos, H. & Habig, K.-H. (2003). *Tribologie-Handbuch*. Vieweg-Verlag, 2nd edition, Wiesbaden, Germany

Grün, F.; Godor, I. & Eichlseder, W. (2007). Schadensorientierte Prüfmethoden und abgeleitete Funktionsmodelle für Gleitwerkstoffe. *Tribologie und Schmierungstechnik*, Vol. 54, No. 5, pp. 26-30

Hofmann, U.; Bögel, A.; Hölzl, H. & Kuhn, H.-A. (2005). Beitrag zur Metallographie von Kupfer und Kupferlegierungen. *Praktische Metallographie*, Vol. 42, No. 7

Kuhn, H.-A.; Hölzl, H.; Hofmann, U.; Kudashov, D. & Zauter, R. (2004). Metallographie heterogener Hochleistungs-Werkstoffe auf Kupferbasis. In: *Fortschritte in der Metallographie (ed. P. Portella), Sonderbände der Praktischen Metallographie*, Vol. 35 (publisher: Petzow, G.), Verlag Werkstoff-Informationsgesellschaft, Oberursel, Germany

Kuhn, H.-A.; Koch, R. & Knab, M. (2007). Thermal stability of lead-free wrought Cu-base alloys for automotive bushings, *World of metallurgy ERZMETALL*, Vol. 60, pp. 199-207

Larson, F.R.; Miller, J. (1952). A time-temperature relationship for rupture and creep stresses. *Trans ASME*, Vol. 74, pp. 765-771

Plewes, J.T. (1975). High strength Cu-Ni-Sn-Alloys by thermo-mechanical processing. *Met. Trans.*, Vol. 6A, pp. 537-544

Pucher, K.; Weber, K.; Huk, V. & Kuhn, H.-A. (2009). *Entwicklung Pb-freier Lagerwerkstoffe für hochbelastete Monometall-Gleitlager*. Tribologie-Fachtagung, Göttingen, Germany, September 21-23

Römer, E. & Bartz, W.J. (1981). *Gleitlagertechnik*. Expert-Verlag, Grafenau/Württ, Germany

Scharf, M. (1999). Copper Alloys in Bearings. In: *Papers of the IWCC Technical Seminar "Copper-the Metal for the New Millennium"*, paper 6

Schrader, A. (1941). *Ätzheft – Anweisung zur Herstellung von Metallschliffen*. Verlag von Gebrüder Bornträger, Berlin, Germany

Weber, K. & Pucher, K. (2009). Neuer Pb-freier Kupferwerkstoff für Gleitlageranwendungen in Verbrennungsmotoren und Getrieben. *Metall*, Vol. 63, No. 11, pp. 564-567

Weber, K. (2009). Kupfer-Zink-Legierung/Verfahren zur Herstellung und Verwendung. *Patent-Offenlegungsschrift DE 2007 029 991*

Zum Gahr, K.H. (1988). Entwicklung und Einsatz verschleißfester Werkstoffe. *Materialwissenschaft und Werkstofftechnik*, Vol. 19, pp. 223-230

Zum Gahr, K.H. (1992). Reibung und Verschleiß. Ursachen-Arten-Mechanismen. In: H. Grewe (publisher): *Reibung und Verschleiß*. Verschleiß-Symposium der DGM, DGM Informationsgesellschaft, Germany, pp. 3-14

5

Mechanical Properties of Copper Processed by Severe Plastic Deformation

Ludvík Kunz
Institute of Physics of Materials,
Academy of Sciences of the Czech Republic
Czech Republic

1. Introduction

Copper has been used for thousands of years and its mechanical properties are well known. Its utilisation in many branches of industry is intensive and has been steadily increasing in recent decades. The major applications are in wires, industrial machinery, copper-based solar power collectors, integrated circuits and generally in electronics. Copper can be also recycled very effectively.

Detailed studies of a relation of mechanical properties and microstructure have been performed in the second half of the last century. The basic data can be found in review papers, (e.g. Murphy, 1981). Cu is a simple f.c.c. material. This is why it has been frequently used as a model material for basic studies of damage mechanisms in metals, particularly fatigue and creep. It belongs from this point of view to the most thoroughly investigated materials. The research was conducted both on polycrystals and single crystals. The great deal of the pool of basic knowledge on fatigue damage mechanisms, changes of dislocation structures, localisation of cyclic plasticity, initiation and propagation of fatigue cracks was acquired just on this model material.

The effort to increase mechanical properties of engineering materials led in the last two decades to application of severe plastic deformation (SPD), resulting in fine-grained structures, exhibiting improved mechanical properties. Naturally, copper was again a suitable material for basic research and, simultaneously, an improvement of its tensile and fatigue strength is a permanent research challenge.

This chapter briefly summarises basic fatigue properties of conventionally grained (CG) copper. However, the main concern is to present and discuss the mechanical behaviour of ultrafine-grained (UFG) Cu prepared by one of the SPD methods, namely by equal channel angular pressing (ECAP). This method enables the production of the UFG material in bulk. The emphasis is put on the fatigue properties and their relation to the UFG microstructure. Discussion of recent, - and, in some cases inconsistent - results on UFG Cu published in literature, is accentuated. This is an issue of the relation of the cyclic softening/hardening to the stability of UFG structure, the influence of mode of fatigue loading on the dynamic grain coarsening, related fatigue life, the mechanism of the cyclic slip localization and initiation of fatigue cracks.

2. Conventionally grained copper

Murphy (1981) summarised in a comprehensive review the basic knowledge acquired until the eighties of the last century. The extensive set of experimental data indicates that the minimum fatigue strength, σ_c, of annealed Cu at 10^9 cycles to failure is 50 MPa. The majority of data published in literature falls into the interval of 50 to 60 MPa. This holds for load symmetrical cycling. The ultimate tensile strength, σ_{UTS}, of investigated coppers was in the range of 200 - 250 MPa, reflecting the wide range of annealing times, temperatures and the source material. The S-N curve of CG Cu of commercial purity 99.98 % with the grain size of 70 μm (determined by mean intercept length) can be well described in the interval form 10^4 to 10^7 cycles by the equation

$$\sigma_a = k_1 N_f^{-b}, \tag{1}$$

where σ_a is the stress amplitude, N_f number of cycles to failure, k_1 = 388 MPa and b = 0.107 (Lukáš & Kunz, 1987). The copper was annealed for 1 hr. in vacuum; its σ_{UTS} was 220 MPa and the yield stress, $\sigma_{0.2}$, was 37 MPa.

Tensile mean stress results in a decrease of fatigue life. At the beginning the expressive decrease of lifetime with increasing tensile mean stress is observed. It is more severe than that predicted by Gerber parabola; however, for higher mean stresses the effect is getting weak. The constant fatigue life curves in the representation σ_a vs. tensile mean stress, σ_{mean}, exhibit a plateau for medium values of σ_{mean}.

The dependence of fatigue strength on the grain size was found to be quite weak. Thompson & Backofen (1971) observed that there is no effect of grain size ranging in the interval 3.4 to 150 μm on the fatigue life in the high-cycle fatigue (HCF) region. This behaviour was attributed to easy cross-slip. Later on the grain size effect was not substantiated even though the grain size was varied from 50 μm to 0.5 mm. Further goal-directed study of the effect of grain size indicated a weak decrease of the fatigue limit expressed in terms of the dependence of N_f on the total plastic strain amplitude, $\varepsilon_{a,tot}$, with increasing grain size (Müllner et al., 1984). The effect increases with decreasing plastic strain amplitude and increasing lifetime. Generally, it can be summarised that the fatigue life curves expressed both as S-N curves or dependences of number of cycles to failure on the total strain amplitude depend on the grain size insignificantly. This holds especially for fatigue limits based on 10^7 cycles (Lukáš & Kunz, 1987).

The Coffin-Manson plot, however, depends strongly on the grain size. It is shifted to lower values of plastic strain amplitude, ε_{ap}, for a given number of cycles to failure with increasing grain size D. Experimental data on the number of cycles to failure for plastic strain amplitude of 1 x 10^{-4} taken form (Lukáš & Klesnil, 1973; Polák & Klesnil, 1984; Kuokkala & Kettunen, 1985) indicates a roughly linear increase of N_f with $D^{-1/2}$. The plastic strain amplitude fatigue limit based on 10^7 cycles was found to be grain size dependent, being 4 x 10^{-5} for fine-grained copper and 2.3 x 10^{-5} for coarse-grained Cu. The explanation is sought in the different conditions for non-propagation/propagation of short cracks, which physically determine the fatigue limit.

Copper exhibits strong work hardening, which is typical for single-phase f.c.c. structures. The tensile strength of annealed material can be increased by 100 % due to 80 % cold working.

Cyclic loading of annealed Cu results in rapid cyclic hardening followed by a long period of cyclic softening. In the HCF region the period of rapid hardening takes only about 1 to 3 % of the total number of cycles to failure. This behaviour is characteristic for broad temperature interval (Lukáš & Kunz, 1988). Fatigue of hardened metals and alloys generally results in cyclic softening. The intensity of this effect depends on the stability of the hardened structure and on the cyclic conditions; i.e., the level of the stress or strain amplitude (Klesnil & Lukáš, 1992). The hardness of Cu, both annealed and cold worked tends to reach during cycling the nearly same value, provided the fatigue life is of the order of 10^7 cycles. The fatigue strength, irrespective of the fatigue hardening followed by softening, has been shown to increase nearly linearly with the σ_{UTS} reached by cold working. This is fulfilled at least up to 40% of cold work. At higher strengths and more severe cold work Murphy (1981) signalises a number of anomalous results, without any detailed explanation. In some cases of high cold work the fatigue strength is even so low as in the case of annealed Cu. This is why for engineering applications it is generally recommended that the σ_{UTS} of unalloyed Cu is restricted to less than 325 MPa, corresponding to ~ 30 % cold work.

The cyclic stress-strain response of Cu can be well described by the cyclic stress-strain curve (CSSC) defined on the basis of the stress and strain amplitudes determined for 50 % of the total number of cycles to failure. For fine-grained copper with the grain size of 70 μm it holds

$$\sigma_a = k_2 \varepsilon_{ap}^n , \qquad (2)$$

where k_2 = 562 MPa and n = 0.205. The CSSC of coarse-grained Cu is shifted to lower plastic strain amplitude values and curve for very large grains exhibits even deviation from the power law in the range of stress amplitudes from about 70 to 100 MPa (Lukáš & Kunz, 1987). This effect is related to the development and increase of volume fraction of persistent slip bands (PSB) in the matrix (Lukáš & Kunz, 1985; Wang & Laird, 1988). The CSSC of Cu single crystals exhibit a plateau (Mughrabi, 1978) which extends over about two decades of plastic shear strain amplitude and which is related to inhomogeneous deformation localised in PSBs.

Decreasing temperature has been known to increase the fatigue strength of both annealed and cold worked Cu. S-N curves are shifted towards higher number of cycles with decreasing temperature. A reduction of temperature results in an increase of the saturation stress amplitude for the same plastic strain amplitude. On the other hand, the Coffin-Manson curves were found to be independent of temperature (Lukáš & Kunz, 1988).

The fatigue behaviour of CG Cu is determined by its dislocation structure, which develops during cyclic loading and which is a function of external loading parameters. The dislocation activity results in a cyclic slip localisation, which manifests itself by development of a surface relief. There is a clear relation between the surface relief and the underlying dislocation PSB structure in CG Cu. The regions of PSBs are characteristic with higher plastic strain amplitudes than the surrounding interior structure. The structural dimensions, i.e. the characteristic dimensions of vein structure, PSBs and cells are generally large compared with the characteristic structural dimensions of UFG Cu. This indicates that the basic knowledge on the cyclic strain localisation and on mechanisms of crack initiation obtained on CG Cu cannot be straightforwardly applied to the UFG structures.

3. Ultrafine-grained copper

Severe plastic deformation of metallic materials has attracted intensive attention of researchers in material science within the last two decades. The main expectation both of research and industry is to improve the mechanical properties of metals and alloys by substantial grain refinement. It has been well known for a long time that a fine-grained material exhibits better strength and hardness than that one which is coarse-grained. Reduction of the grain size usually also improves fracture toughness. The physical reason of improved mechanical properties lies in the higher grain boundary volume in fine-grained structures, which makes the dislocation motion and resulting plastic deformation more difficult. For many materials the yield stress follows the Hall-Petch equation in a very broad range of grain size between 1 μm and 1 mm (Saada, 2005). Deviations from this law are observed only for very coarse grained and for nano-grained structures.

3.1 Equal channel angular pressing

Equal channel angular pressing is the most popular SPD technique. There are number of papers describing the fundamentals of this process and material flow during pressing, e.g. (Segal, 1995; Valiev & Langdon, 2006). The principle of the method is very simple. It consists of pressing of a rod-shaped billet through a die with an angular channel having an angle Φ, often equal to 90°, Fig. 1. A shear strain is introduced when the billet pressed by a plunger passes through the knee of the channel. Since the cross-sectional dimensions of the billet remain constant after passing the channel, the procedure can be repeated. The result is a very high plastic strain of the processed material. The majority of laboratory ECAP dies has a channel with a quadratic cross-section. The billets of corresponding dimensions can be rotated by increments of 90 degree between particular passes. Indeed, the rotating procedure is feasible also for dies and samples with circular cross-section. Four different ECAP routes are distinguished. Route A means repetition of pressing without any billet rotation. Route B_A represents rotation of the billet by 90° in alternating directions, route Bc means rotation by 90° in the same direction after each consecutive pressing and the route C represents the rotation by an angle of 180° after each pass.

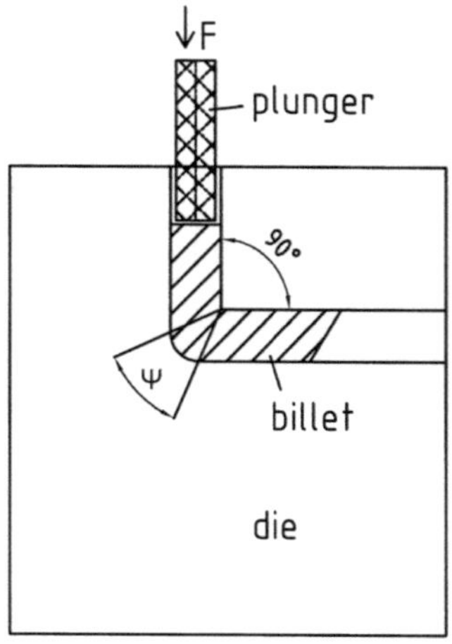

Fig. 1. Principle of ECAP procedure

The equivalent strain, ε, reached by pressing through a die characterised by outer arc Ψ of curvature of the cannels inclined mutually at an angle Φ, is given by the relationship (Valiev & Langdon, 2006)

$$\varepsilon = \left(N/\sqrt{3}\right)\left[2\cot\left\{(\Phi/2)+(\Psi/2)\right\}+\Psi\,\mathrm{cosec}\left\{(\Phi/2)+(\Psi/2)\right\}\right]. \tag{3}$$

Processing by SPD methods and the investigation of the resulting UFG structures and their properties is a mater of rapidly increasing number of research papers. The current results are regularly presented at the NanoSPD conference series; the last was held in 2011 (Wang, et al., Eds, 2011). A plenty of improvements of ECAP procedure has been proposed in the past, e.g. a rotary-die putting away the reinserting of the billet into the die, dies with parallel channels or the application of back pressure. Though the requirements of process improvements and economically feasible production of UFG materials in sufficient volumes activates development of plenty of SPD methods like accumulative roll bonding, multiaxial forging or twist extrusions, the majority of basic knowledge on UFG materials is based just on research on materials processed by simple ECAP. This holds also for Cu, which was as the simple f.c.c. model material used for pioneering studies on fatigue behaviour of UFG structures prepared by SPD (Vinogradov et al., 1997; Agnew et al., 1999).

3.2 Microstructure

The grain size of UFG materials is typically in the range of 10^2 to 10^3 nm. This is a transition region between the CG materials and nanostructured metals, where the grain boundaries play a decisive role during plastic deformation. Quantitative determination of grain or cell size of materials processed by ECAP is often complicated by the fact that the size is varying broadly between hundreds of nanometres and some micrometers and by not well-defined boundaries in TEM images (e.g. Vinogradov & Hashimoto, 2001). Experimental study of evolution of microstructure in Cu by Mishra et al. (2005) shows that the first few ECAP passes result in an effective grain refinement taking place in successive stages: homogeneous dislocation distribution, formation of elongated sub-cells, formation of elongated subgrains and their following break-up into equiaxed units, while the microstructure tends to be more equiaxed as the number of passes increases. Later on, the sharpening of grain boundaries and final equiaxed ultrafine grain structure develops.

The microstructure of Cu prepared by ECAP can vary in many parameters. This is an essential difference to CG Cu. UFG Cu can differ in the grain size distribution, shape and orientation, dislocation structure and dislocation arrangement in grain boundaries, in texture and misorientation between adjacent grains, which reflects the details and conditions of the ECAP procedure (Valiev et al., 2000; Zhu & Langdon, 2004). The mutual orientation of structural units cannot be satisfactorily described as high-angle random orientation. There are regions where low angle boundaries are present, and also regions which can be described as regions of near-by oriented grains. That is why instead of the term "grain size" a term "dislocation cell size" is sometimes used.

An example of a UFG structure of Cu of purity 99.9 % is shown in Fig. 2. Cylindrical billets of 20 mm in diameter and 120 mm in length were produced by eight ECAP passes by the route Bc. After the last pass through the die the samples of 16 mm in diameter and 100 mm in length were machined from the billets. The severe plastic deformation was conducted at room temperature. The microstructure as observed by transmission electron microscopy (TEM) in the middle of a longitudinal section of the cylindrical sample is shown in Fig. 2a. The structure in transversal direction is shown in Fig. 2b. The average grain size, determined on at least 10 electron micrographs, is 300 nm. This is in full correspondence with the results by Mingler et al. (2001). They report that ECAP of pure Cu leads to the most frequent grain size of 300 nm irrespective to the number of passes applied. The size distribution, however, is getting narrower with increasing number of passes and both total and grain-to-grain misorientation tends to reach high angle type. Similarly, the grain size of ~ 200 nm was reported by Agnew et al. (1999), or grain size in the range from 100 to 300 nm by Besterci et al. (2006); however, here in dependence on the details of the ECAP procedure. Vinogradov & Hashimoto (2001) distinguish between the structures with different morphological features: the equiaxial structure, referred to as "A", and elongated grain structure called "B". They note that in the course of the ECAP procedure, it is highly possible to obtain a mixture of the type A and B structures. The microstructure in Fig. 2a resembles the type B and the structure in Fig. 2b the equiaxial type A.

Fig. 2a. Microstructure of Cu after ECAP as observed in TEM, longitudinal section

Fig. 2b. Microstructure of Cu after ECAP as observed in TEM, transversal section

For characterisation of microstructure, electron back scattering diffraction (EBSD) has been recently used beyond the TEM (Wilkinson & Hirch, 1997). EBSD analysis is predominantly focused on the experimental determination of misorientation of a crystallographic lattice between adjacent analysed points. This technique, in contrast to TEM of thin foils, enables one to investigate the changes of microstructure in the course of fatigue loading, provided the same area of the specimen gauge length is examined. An example of a microstructure as observed by EBSD and the analysis of EBSD data is shown in Fig. 3. A grain map is shown in Fig. 3a. Grains, defined as areas having the mutual misorientation higher than 5 degrees (threshold angle, which can be adjusted), are marked by particular colours. Fig. 3b brings the grain size distribution. Indeed, the direct comparison with the structure displayed by TEM is not possible, because the evaluation procedure of back scattered electron diffraction images is primarily dependent on the adjusted threshold angle. However, for the constant threshold angle the method enables detection of changes of microstructure and grain orientation due to fatigue loading or temperature exposition.

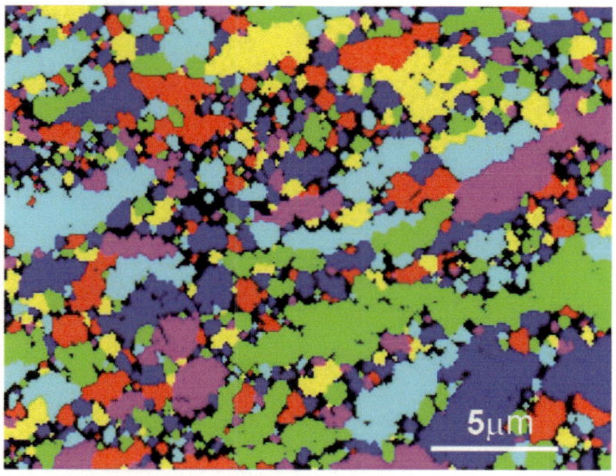

Fig. 3a. Grain map of UFG Cu as displayed by EBSD

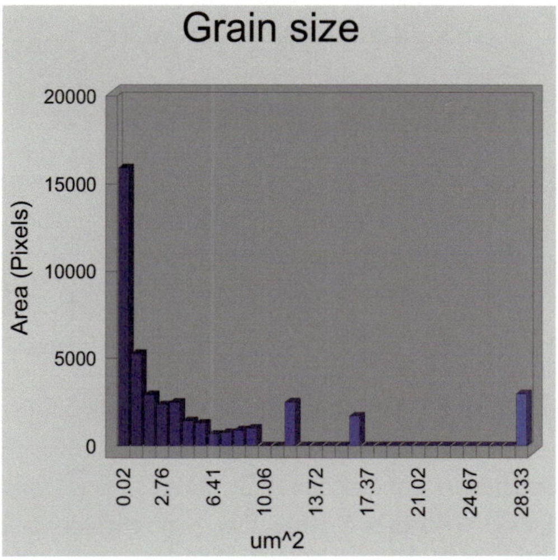

Total grains: 912; Average size: 0.45; Average ASTM: 18.8; Threshold angle: 5°

Fig. 3b. Analysis of the grain size and grain size distribution

The inhomogeneity of severe plastic deformation by ECAP is documented on UFG Cu of very high purity of 99.9998 % in Fig. 4. The overetched surface of material exhibits traces of non-uniform deformation during SPD. The simple sketches in Fig. 4a, based on the appearance of the surface markings, highlight the mutual shift of the layers 1 and 2 on both sides of the band in between. The material in both layers appears not to be sheared during the last ECAP paths so intensively as the material in the bands. This observation implies inhomogeneity of shearing during ECAP by the route C. Similar surface observation after processing by the route B_C is shown in Fig. 4b. The structure exhibits traces of shearing on two slip systems corresponding to the billet rotation.

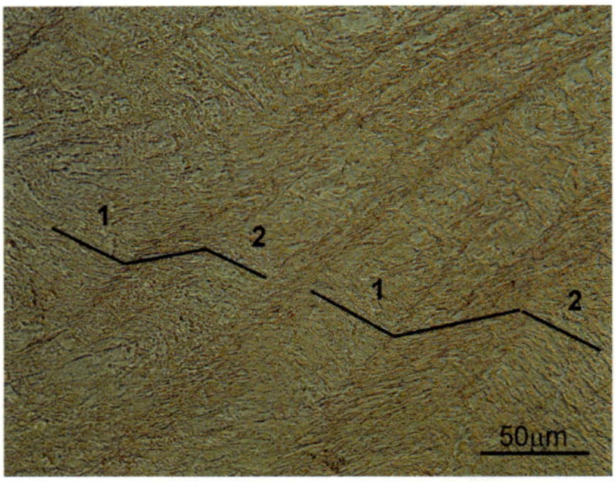

Fig. 4a. Severely etched surface of UFG Cu prepared by ECAP, route C

Fig. 4b. Severely etched surface of UFG Cu prepared by ECAP, route Bc

The UFG structure produced by ECAP is sensitive to the technological details of the process, lubrication, deformation rate, dimensions of the die etc. No doubt these factors influence the microstructure and finally the properties of UFG material. So, the diversity in behaviour of "nominally" identical UFG structures produced in different laboratories makes the comparison of results published in literature troublesome. A variety of possible structures give rise to significant scattering of experimental data on fatigue behaviour published to date (Vinogradov & Hashimoto, 2001).

3.3 Tensile properties

The stress-strain diagrams of UFG Cu prepared by eight passes by route Bc and specified in the preceding paragraph are shown in Fig. 5. The diameter of the gauge length of specimens was 5 mm. Results for three values of the loading rate, 1, 10 and 100 mms^{-1} are presented. Obviously, there is no strain rate influence in the range of rates applied. All three curves are located in a narrow interval. The differences between them are smaller than the difference between two specimens tested at the same strain rate 1 mms^{-1}. The shape of the curves is identical with that observed for UFG Cu prepared by 10 passes by the route C (Besterci et al., 2006). The curves exhibit a long elastic part at the beginning. The basic tensile properties determined as an average of four measurements are given in Tab.1.

ultimate tensile strength σ_{UTS}	yield stress $\sigma_{0.1}$	yield stress $\sigma_{0.2}$	modulus of elasticity E
[MPa]	[MPa]	[MPa]	[GPa]
387 ± 5	349 ± 4	375 ± 4	115 ± 11

Table 1. Tensile properties of UFG Cu prepared by ECAP, route Bc, 8 passes

The ultimate tensile strength after 8 passes, σ_{UTS}, is 387 MPa. The yield strength $\sigma_{0.2}$ = 375 MPa is very close to the σ_{UTS} and makes 97 % of it. The scatter of the data determined on particular specimens is quite small.

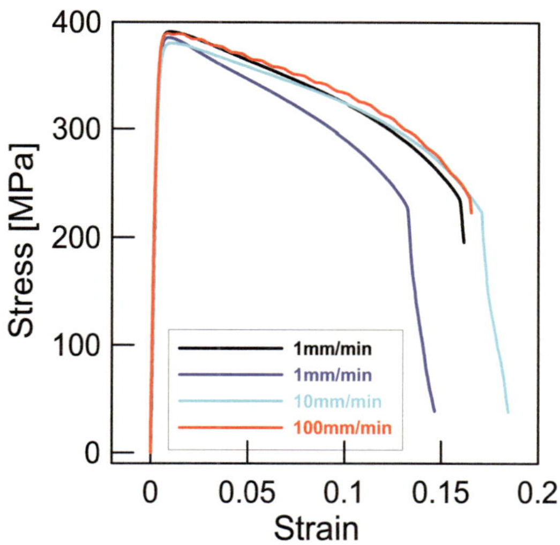

Fig. 5. Tensile diagrams of UFG Cu prepared by ECAP, route Bc

The basic tensile data reported in literature for Cu processed by ECAP in different laboratories differs considerably. For instance, Besterci et al. (2006) report σ_{UTS} values ranging from 410 to 470 MPa for number of passes between 3 and 10, and Goto et al. (2009) 443 MPa for 12 passes by the processing route Bc. Vinogradov et al. (2001) report the value ~520 MPa for Cu processed by 12 Bc passes. Generally, during the first passes a rapid increase of strength is observed. However, later on, the strength saturates or even decreases.

From the comparison of tensile diagrams of CG and UFG Cu, Fig. 6, it can be seen that there is a substantial difference in the initial part of the diagram, indicating very high yield strength of UFG Cu and very low one for CG material.

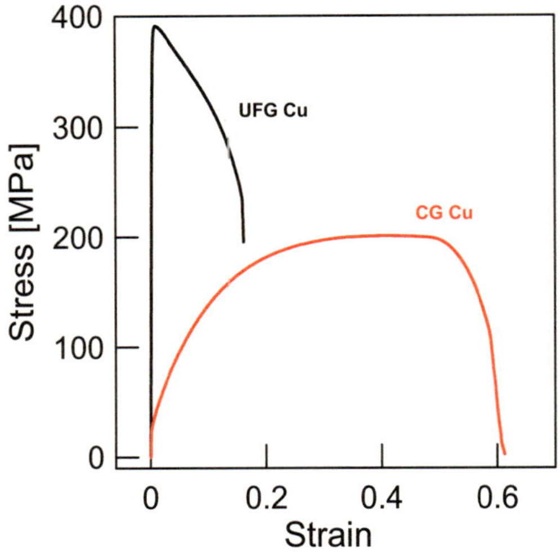

Fig. 6. Comparison of tensile diagrams of UFG and CG Cu

3.4 Fatigue strength

The improvement of fatigue performance of Cu by ECAP processing was experimentally documented in, for example, (Agnew et al., 1999; Vinogradov, & Hashimoto, 2001; Höppel, & Valiev, 2002; Kunz et al., 2006 and Höppel et al., 2009). The S-N curve of Cu prepared by eight ECAP passes by the route Bc, having the tensile properties given in Tab.1 and the structure shown in Figs. 2 and 3 is shown in Fig. 7. The fatigue loading was performed in load symmetrical cycle in tension-compression. The number of cycles to failure increases continuously with decreasing stress amplitude in very broad interval ranging from low-cycle fatigue (LCF) region up to the gigacycle region. Arrows indicate the run-out specimens. The experimental points in the interval of numbers of cycles to failure spreading over 7 orders of magnitude cannot be well approximated by a straight line in log-log representation. The description by power law is, however, possible in the first approximation in the shorter interval, namely from 10^4 to 10^8 cycles by the equation (1) with constants k_1 = 584 MPa and b = -0.078. The fatigue life of UFG Cu is substantially higher than that of annealed CG Cu and also than that of cold worked copper reported by Murphy (1981). The S-N curve for UFG material is shifted by a factor of 1.7 towards higher stress amplitudes for a given number of cycles to failure when compared to the cold worked material.

The UFG data in Fig. 7 shows the results of experiments on identical material but performed on different fatigue machines: servohydraulic, resonant and ultrasonic; and on different types of specimens. The frequency of loading with a sine wave was in the interval of 1 Hz to 20 kHz. Similarly to the CG Cu there is no apparent influence of loading frequency on the S-N curve. All experimental data fall into one scatter band.

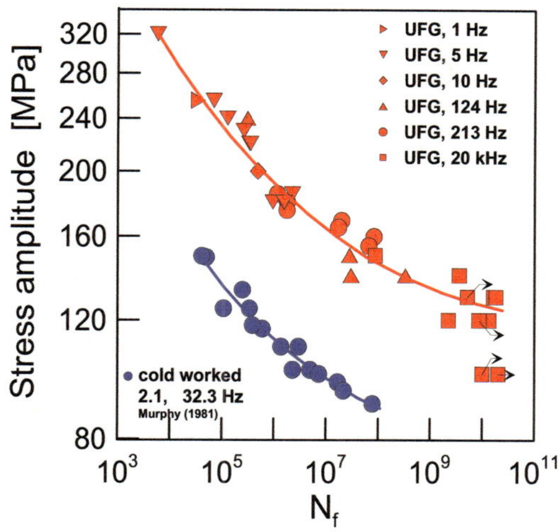

Fig. 7. Comparison of S-N data for UFG and cold worked Cu

From Fig. 7 it follows that the fatigue limit of UFG Cu based on 10^8 cycles is 150 MPa. The σ_{UTS} of this copper is 387 MPa, see Tab.1. Fig. 8, which is redrawn from (Murphy, 1981), shows the relation between the fatigue limit (on the basis of 10^8 cycles) and σ_{UTS} for oxygen free cold worked Cu of purity higher than 99.99 %. Increasing tensile strength by cold work increases the fatigue limit. This reasonably holds for σ_{UTS} up to ~350 MPa. At higher tensile strength, related to cold reduction above ~40 %, a large scatter of data exists. In some cases even a decrease of fatigue limit down to the annealed Cu was observed. The experimental point corresponding to severely deformed UFG Cu, which is shown in Fig. 8 by the full symbol, is situated on the right-hand side of the scatter-band of data. The result qualitatively fits into the general trend of increasing fatigue limit with ultimate tensile strength.

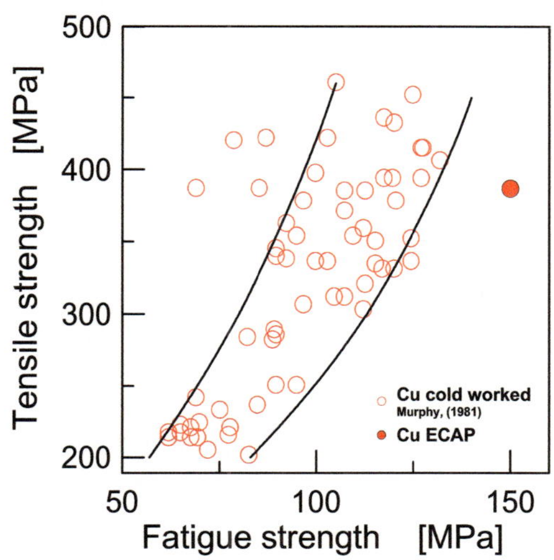

Fig. 8. Relation of tensile and fatigue strength for fatigue limit based on 10^8 cycles for cold worked Cu (Murphy, 1981) and UFG Cu

From the recent overview of the cyclic deformation and fatigue properties of UFG materials by Mughrabi & Höppel (2010) it arises that the fatigue behaviour depends strongly on parameters of the ECAP procedure, purity of material and type of fatigue loading. The discussion, interpretation and, in particular, the comparison of results published in literature, requires all the details of the UFG structures produced in different laboratories and also the external loading parameters and conditions to be taken into account.

In the early studies it has been experimentally shown that the equiaxial grain structure of the lamellar-like type B lasts longer under the same stress amplitude than the equiaxial type A (Vinogradov & Hashimoto, 2001). Similar observations were made also on other materials like Ti alloys; however, the available knowledge is not enough to declare that the lamellar-like structures of UFG Cu are generally better than equiaxial ones.

The experimental data presented in Fig. 7 gives evidence that the UFG structure of Cu can exhibit substantially better fatigue strength expressed in terms of S-N curves than the CG Cu. However, there is also data in literature that indicates quite poor or no improvement of fatigue strength in the high-cycle fatigue HCF region. Han et al. (2007; 2009) and Goto et al. (2008) observed the strong enhancement of fatigue life in LCF range but very weak effect in long-life regime. The fatigue strength of 99.99 wt% Cu processed by four passes by route Bc coincided with that of fully annealed copper for 3 x 10^7 cycles. This fatigue strength was only slightly enhanced by an increase in the number of ECAP passes and by a decrease in purity. The σ_{UTS} of coppers investigated in these studies was high. The corresponding points [σ_{UTS}, fatigue strength for 10^8 cycles] would lie on the opposite side of the scatter band in the Fig. 8 than the full point characterising the properties of Cu having the S-N curve shown in Fig. 7. It indicates that the large scatter of data reported by Murphy (1981) for cold worked Cu is relevant also to the severely plastically deformed Cu.

Fig. 9 compiles the available majority published experimental results up to now on of fatigue life of UFG Cu prepared by ECAP cycled under constant stress amplitudes. S-N data was obtained in different laboratories. It is remarkable that the field of the S-N points splits up into two distinct bands. The inspection of the legend in the figure shows that the material purity could be a parameter influencing the HCF strength. The band A covers S-N points for low purity UFG coppers (purity in the range from 99.5 and 99.9 %), while the band B covers S-N points for high purity UFG coppers (purity in the range from 99.96 to 99.9998 %). The details of the ECAP process, particularly the type of paths (Bc or C), seem to have only minor effect on fatigue performance. The bands merge into one in the LCF region and obviously diverge in the HCF region. The average stress amplitude corresponding to the 10^7 cycles to failure is around 160 MPa for band A and around 90 MPa for band B. The trend of the bands indicates that the most pronounced effect of purity can be expected in the very high-cycle fatigue (VHCF) region.

The S-N curves of two coppers of substantially different purity tested in a goal-directed research are shown in Fig. 10. Cu was processed by two ECAP routes, namely Bc and C. The fatigue tests were carried out in one laboratory under the same testing conditions. Thus the effect of variances in testing procedures (except of the different specimen shape) is eliminated. It can be seen that the fatigue strength of high purity copper is lower than that of low purity copper. The figure also shows that the ECAP route affects the fatigue strength of pure material. Both the effects, i.e. purity and route are more pronounced for low stress amplitudes. At high stress amplitudes corresponding to lifetimes below 10^4 cycles the effects are practically wiped off.

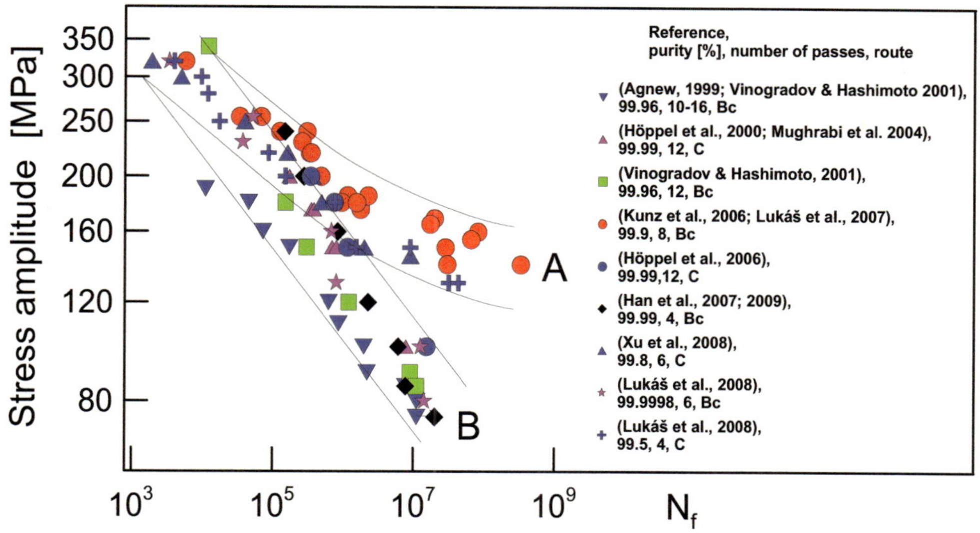

Fig. 9. S-N data of UFG Cu of different purity and processed by different routes and number of ECAP passes

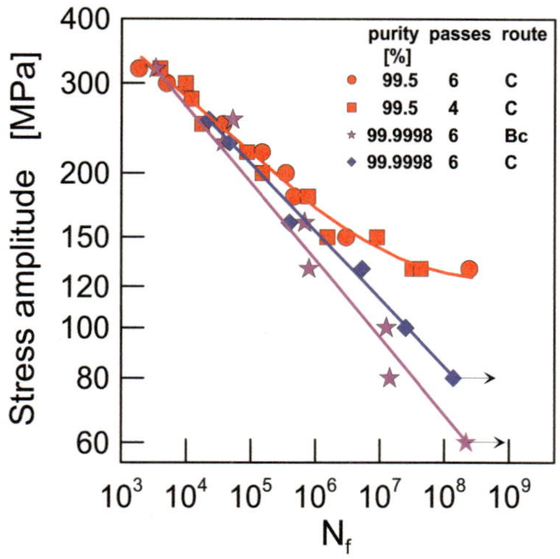

Fig. 10. Influence of purity on S-N curves of UFG Cu

The explanation of the large differences in the fatigue resistance of UFG Cu in the HCF region shown in Fig. 9 can be sought either in the stability of the microstructure during fatigue loading or in the mechanism of the strain localisation and in the fatigue crack initiation. The stability of UFG structure and crack initiation will be discussed later in paragraphs 3.6 and 3.7.

Decreasing temperature results in higher fatigue resistance of UFG Cu. Fig. 11 compares the fatigue lifetime at RT and at temperature of 173 K. The low temperature S-N curve is clearly shifted towards higher stress amplitudes. The shift in the interval from 10^4 to 10^7 cycles to failure is of about 40 MPa.

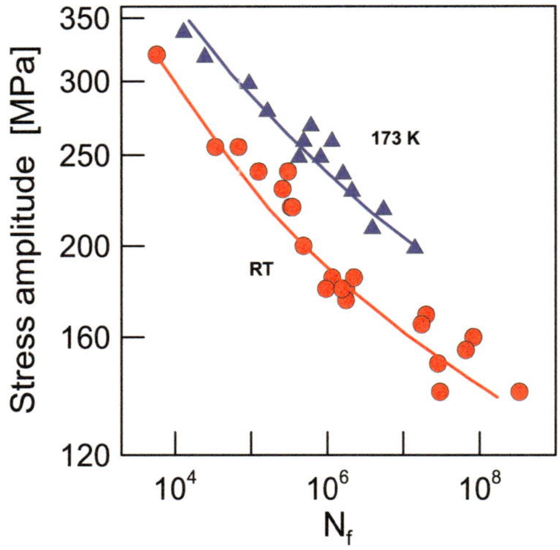

Fig. 11. Influence of temperature on S-N curve of UFG Cu

The influence of mean stress on fatigue life is shown in Fig. 12. The tensile mean stress of 200 MPa decreases the life by a factor of one and half of the magnitude. This holds from the HCF region up to the fatigue live of 10^5 cycles, which corresponds to the stress amplitude of 180 MPa. The maximum stress in cycle with the stress amplitude 180 MPa is 380 MPa, which is very close to the σ_{UTS} = 387 MPa. It is interesting that under these conditions, when the maximum stress in a cycle nearly touches the tensile strength of the material, the fatigue life is still very high. The next small increase of the stress amplitude means that the tensile strength is exceeded and the fatigue life is getting very short. The scatter of lives at the stress amplitude 190 MPa (horizontal dashed line in Fig. 12) is related to the inherent scatter of tensile strength of particular specimens. Fig. 12 demonstrates that in UFG Cu, which is cycled under stress-controlled conditions, the low-cycle fatigue region is missing.

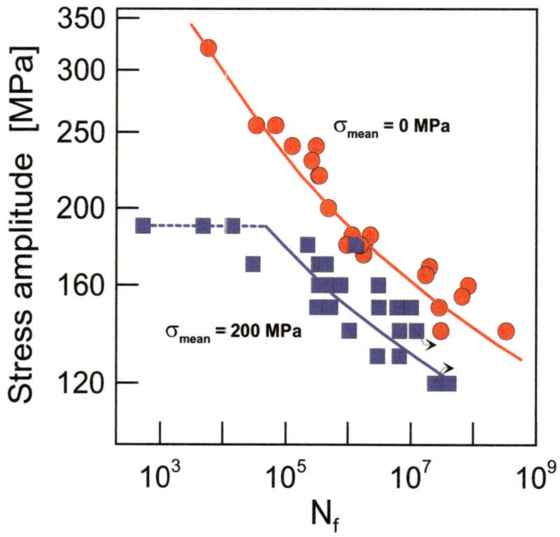

Fig. 12. Influence of mean stress on fatigue life

The transition from the full curve to the horizontal dashed line for the mean stress in Fig. 12 is related to the change of mechanism of failure from fatigue to ductile. The cyclic creep curve, i.e. the development of unidirectional creep deformation during cycling, is shown in Fig. 13 for the specimen with the shortest fatigue life of 528 cycles, Fig. 12. The cyclic creep curve exhibits the typical three stages. The first stage with rapidly decreasing cyclic creep rate is related to the cyclic hardening. The second stage, characterising the decisive part of the fatigue life, is characteristic by a nearly constant cyclic creep rate. The third stage is related to the development of a neck on the specimen, decrease of the specimen cross-section, increase of true stress and final ductile fracture. The fracture surface of the specimen, which failed after 528 cycles at the stress amplitude of 190 MPa, can be seen in Fig. 14. The fracture surface is of a ductile cone and cup type. The fracture surface of a specimen cycled with the stress amplitude 180 MPa, which failed after 1.34 x 10^6 cycles, is shown in Fig. 15. The fatigue crack initiated at the specimen surface. The fracture surface produced by propagating fatigue crack makes only a small part of the final fracture. The majority of the fracture surface is of a ductile type.

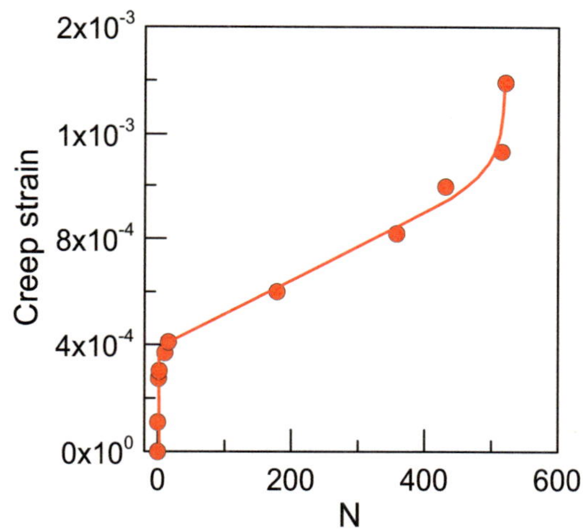

Fig. 13. Cyclic creep curve of UFG Cu, stress amplitude 190 MPa, mean stress 200 MPa

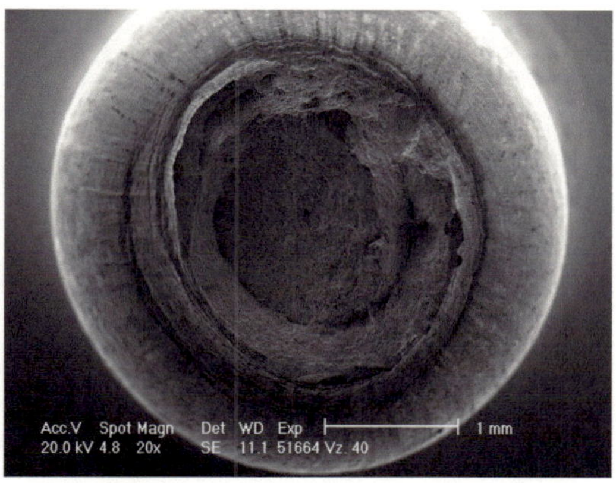

Fig. 14. Fracture surface of a specimen cycled at mean stress of 200 MPa and stress amplitude of 190 MPa

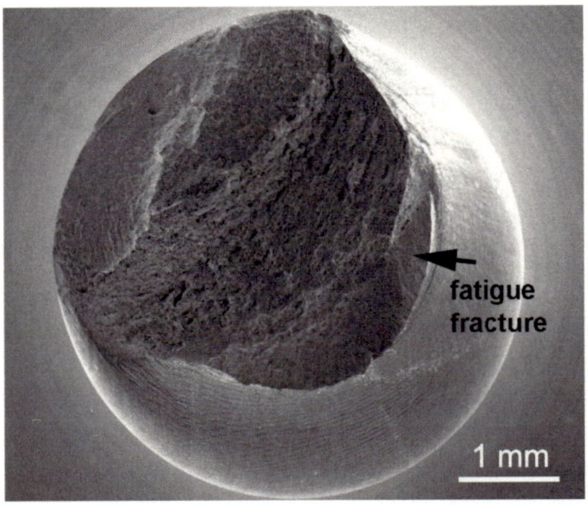

Fig. 15. Fracture surface of a specimen cycled at mean stress of 200 MPa and stress amplitude of 180 MPa

The fatigue lives of UFG Cu are generally higher than those of CG Cu when the comparison is made on the basis of S-N plots, (e.g. Mughrabi, 2004). Just the opposite, however, arises from the comparison on the basis of plastic strain amplitudes. Results of strain-controlled fatigue tests expressed as Coffin-Manson plots show shorter lifetime of UFG than CG Cu (Agnew, 1998, 1999; Vinogradov & Hashimoto, 2001; Höppel, 2006; Mughrabi, 2006). The effect is more pronounced at higher plastic strain amplitudes. This result seems to be obvious, because the UFG Cu is harder but less ductile than CG Cu. Based on these facts, Mughrabi and Höppel (2010) explain it schematically on the total strain life diagrams of UFG and CG materials.

Comparison of Coffin-Manson curves for CG and UFG Cu is shown in Fig. 16. The values of the plastic strain amplitude were obtained from the total plastic strain amplitude controlled tests of CG Cu (Lukáš & Kunz, 1987) and from the stress-controlled tests of UFG Cu (Lukáš et al., 2009). The plastic strain amplitude was determined for a number of cycles equal to ½

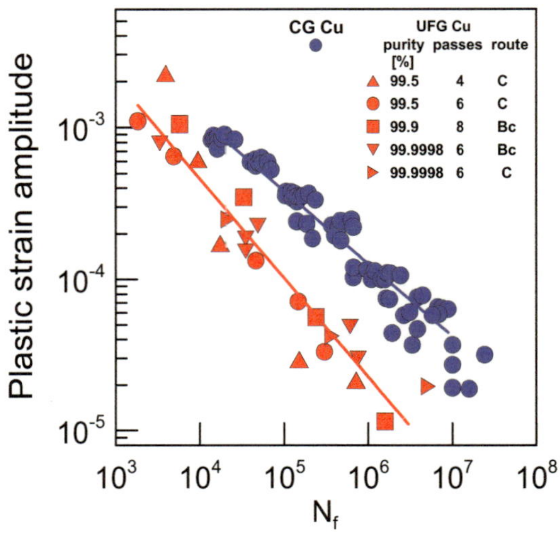

Fig. 16. Comparison of Coffin-Manson curves for CG an UFG Cu

of the total number of cycles to failure. The grain size of CG Cu was 70 μm, and the grain size of UFG Cu was 300 nm. The data for UFG Cu corresponds to different purities, ECAP routes and number of passes. All experimental points are located in one scatter band. This fact means that the plastic strain amplitude could be taken as a unifying parameter for lifetime prediction of UFG Cu.

The experimental fact that the fatigue resistance of UFG Cu is lower than that of CG Cu when loaded at the same plastic strain amplitude is in broad agreement with the expectations according to the total strain fatigue life diagram (Mughrabi & Höppel, 2001).

3.5 Cyclic stress-strain response

3.5.1 Hardening/softening

Cyclic stress-strain response of UFG Cu is presented in Fig. 17. The examples for four stress amplitudes were selected from the set of data obtained by the determination of the S-N points shown in Fig. 7. The relative number of cycles to failure, *N/Nf*, is plotted in dependence on the plastic strain amplitude, ε_{ap}. The tests were performed at constant stress amplitude and were run up to the final failure. It can be clearly seen that the specimens cycled at higher stress amplitudes soften. At the very beginning of the tests, at higher stress amplitudes a quick hardening was observed. This is more easily visible in Fig. 18, which displays the hardening/softening curves in log-log coordinates. For medium stress amplitudes resulting in the lifetime of the order of 10^5 cycles stable stress-strain behaviour can be observed. For small stress amplitudes in HCF region continuous cyclic hardening is a characteristic feature.

Cyclic softening was already observed in the early studies. Agnew & Weertman (1998) reported cyclic softening under controlled total strain amplitude loading in the range of 1×10^{-2} to 5×10^{-4}. The softening was explained by a general decrease of defect density and due to changes of grain boundary misorientation. The effect of softening decreases with decreasing plastic strain amplitude. No softening was observed at $\varepsilon_{ap} < 10^{-3}$. Some light hardening was noticed on the early stage of straining in (Vinogradov et al., 1997; Vinogradov & Hashimoto, 2002), which is in agreement with results presented in Figs. 17 and 18.

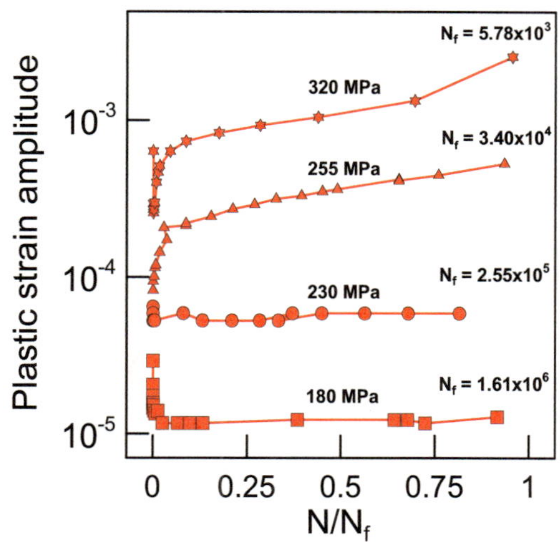

Fig. 17. Dependence of plastic strain amplitude on relative number of cycles to failure N/N_f

The softening process in UFG Cu depends both on the loading and microstructural parameters. The microstructure composed of nearly equiaxed grains with a mean size 200 - 250 μm, type A, exhibits nearly stable cyclic behaviour when cycled at $\varepsilon_{ap} = 10^{-3}$, whereas the elongated structure B exhibits softening under the cycling with the same ε_{ap} (Agnew, 1999; Hashimoto et al., 1999).

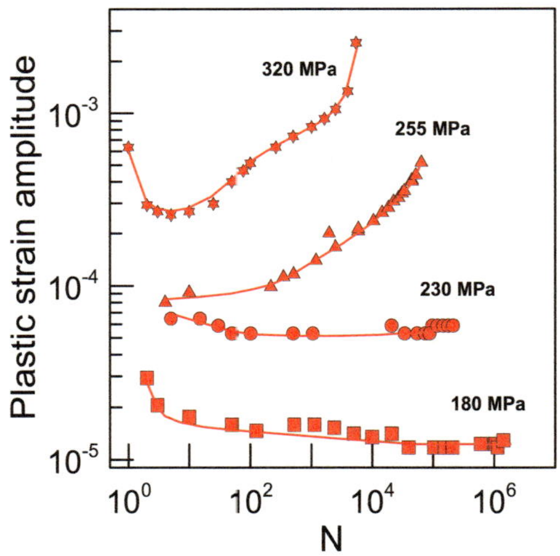

Fig. 18. Cyclic softening/hardening curves of UFG Cu cycled under stress control

More or less severe cyclic softening has been reported to occur at strain-controlled tests. Two curves corresponding to tests of UFG Cu with plastic strain amplitudes of 0.1 % and 0.05 % are shown in Fig. 19. The copper was identical with that used for the determination of cyclic hardening/softening curves under stress control, Figs. 17 and 18. Continuous softening is observed since the beginning of the test. The rapid decrease of both curves for the stress amplitude below ~ 250 MPa is related to the propagation of a magistral fatigue crack through the specimen cross-section.

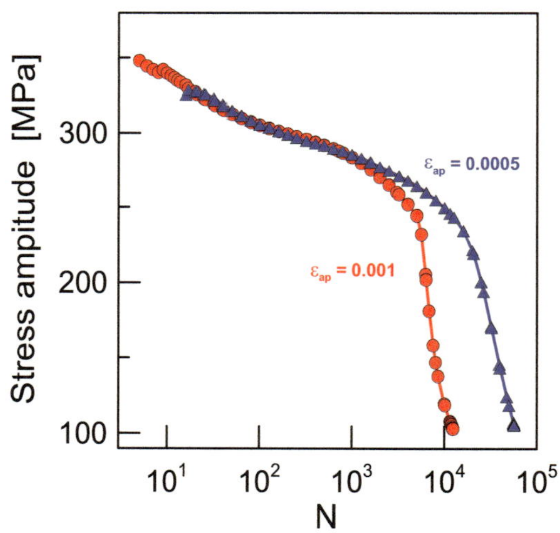

Fig. 19. Cyclic softening curves of UFG Cu cycled under plastic strain control

The softening (and lower fatigue resistance under strain-controlled testing) is often discussed in relation to the low stability of structure produced by SPD. The phenomenon of softening is basically not unexpected, because the UFG microstructure is severely deformed and there is a high stored energy. The mechanism of dynamic grain coarsening in Cu is currently discussed in literature and there is no integrated opinion on this effect.

The cyclic hardening/softening effect in UFG Cu is observed also at low temperatures. An example of the dependence of plastic strain amplitude on number of cycles for stress-controlled fatigue loading at 173 K is presented in Fig. 20. The copper is the same as that one used for determination of S-N curve in Fig. 7 and the hardening/softening curves, Figs. 17 - 19. From the comparison of Figs. 18 and 20 it follows that the characteristic behaviour; i.e., pronounced softening at high stress amplitudes and hardening in HCF region, is qualitatively not influenced by the decrease of temperature.

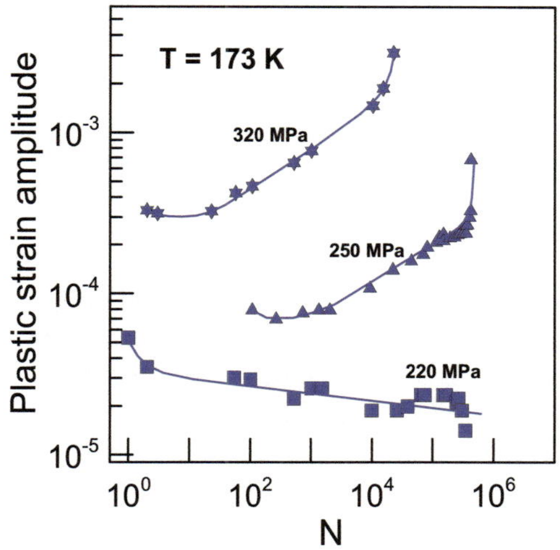

Fig. 20. Cyclic softening/hardening curves of UFG Cu loaded at temperature 173 K

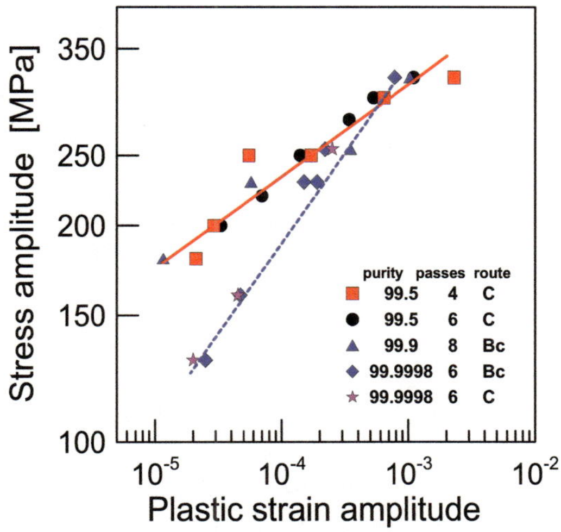

Fig. 21. Influence of purity and ECAP details on cyclic stress-strain curve

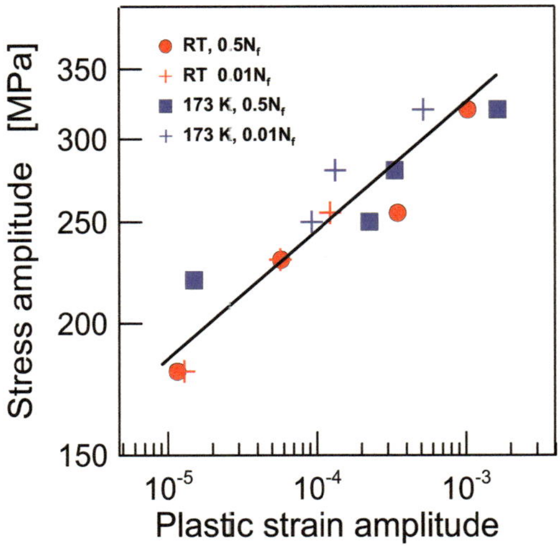

Fig. 22. Influence of temperature and way of determination on cyclic stress-strain curve

3.5.2 Cyclic stress-strain curve

The hardening/softening curves do not generally exhibit saturation behaviour. Pronounced cyclic softening is characteristic in the LCF region whereas weak cyclic hardening is typical for the HCF region. Hence, the CSSC cannot be constructed on the basis of saturated values of stress and strain amplitudes, which characterise the cyclic stress-strain response for the decisive part of fatigue life of cyclically stable materials. For the determination of the CSSC, some convention has to be adopted. One of the often used procedures is to define the CSSC on the stress and strain values corresponding to one half of the fatigue life. The experimental data for UFG Cu of different purity and processed by different ECAP routes by different numbers of passes through the die are shown in Fig. 21. The data points can be separated into two groups according to the copper purity. The curve corresponding to low purity is above the CSSC of high purity. The difference is largest for smallest amplitudes and wanes towards LCF region. No measurable influence of the ECAP processing route and number of passes follows from the presented data.

The influence of temperature and the way of the determination of CSSC of low purity UFG Cu is shown in Fig. 22. The fatigue tests were conducted under controlled stress amplitude. For experiments the same Cu on which was determined the S-N curve shown in Fig. 7 was used. It can be seen that the CSSC is not measurably influenced when the temperature decreases from RT down to 173 K. Furthermore, the experimental points of the CSSC fall into one scatter band when the plastic strain amplitude is conventionally determined for ½ or for 10 % of the fatigue life.

It has been mentioned previously that the fatigue behaviour of UFG Cu depends on details of microstructure and cyclic loading. This fact makes the comparison of data obtained in different laboratories difficult. Fig. 23 presents the comparison of CSSC characterising the behaviour of UFG Cu (which S-N curve is shown in Fig. 7 and structure in Fig.2) and copper investigated in (Höppel et al., 2009; Mugrhabi & Höppel, 2001). It can be seen that there is large discrepancy between both curves. It is interesting to note that in both cases the material under investigation was UFG Cu of low purity, namely 99.9 % processed in a nominally identical way. Detailed comparison of results shows also that the S-N curves of both coppers differ. One of them belongs to the band A in Fig. 9 and the other, published in (Höppel et al., 2009; Mugrhabi & Höppel, 2001) to the band B. For comparison, Fig. 23 also shows the CSSC of CG Cu described by eq. (2) with constants k_2 = 562 MPa and n = 0.205. This curve is well below both the curves characterising the UFG Cu.

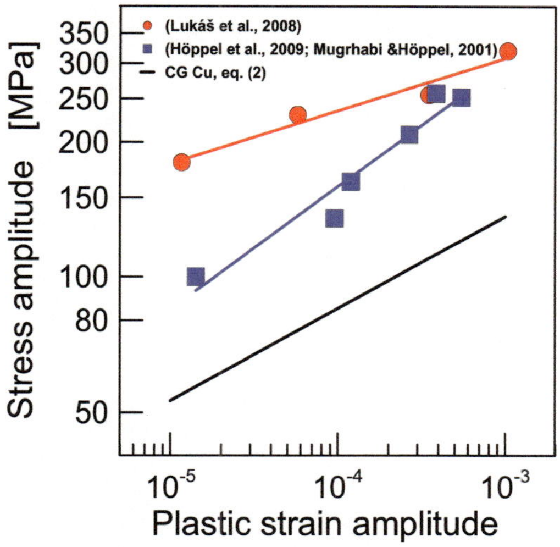

Fig. 23. Comparison of CSSC of two UFG coppers and CG Cu

Fig. 24. Microstructure of UFG Cu after stress-controlled fatigue, stress amplitude 255 MPa

3.6 Stability of UFG structure

Stability of a severely deformed structure is of utmost importance from the point of view of its fatigue properties (Kwan & Wang, 2011). The critical issue of successful application of UFG materials is the long-term stability of microstructure in service where cyclic loads, often with mean stress, are frequent. Also loading at elevated temperatures can be expected in engineering practice. Despite this, knowledge of the stability of UFG structure under dynamic and temperature loading is quite scarce. There are open questions concerning the mechanisms of the grain coarsening both under cyclic loading and temperature exposition.

Fig. 25. Microstructure after 10^{10} cycles

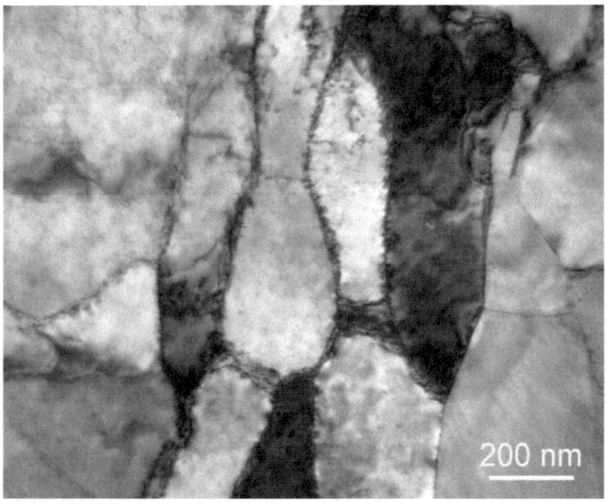

Fig. 26. Microstructure after fatigue loading exhibiting "shaken down" features

ECAP results in structures, which are in metastable sate. There is a natural tendency for recovery and recrystallisation powered by a decrease of high stored energy. Hence, substantial changes of microstructure can be expected in course of fatigue. Really, the total strain-controlled tests of UFG Cu showed a marked heterogeneity of dislocation structure after fatigue loading, which resulted in failure of specimens after 10^4 cycles (Agnew & Weertman, 1998). Three types of structures were described: a) subgrain/cell structure, which resembles the well-known structure from LCF tests of CG Cu; b) a fine-grained lamellar structure as observed in Cu after ECAP; c) areas with large grains with primary dipolar dislocation walls. The first two types of structure were found to make up the majority. Höppel et al. (2002) observed that the intensity of grain coarsening decreases with decreasing plastic strain amplitude and increasing strain rate in a plastic strain controlled test. Pronounced local coarsening of microstructure when compared to the initial state was found after cycling with ε_{ap} = 1 x 10^{-4}. Observation by TEM revealed very pronounced fatigue-induced grain coarsening that occurred in some areas by dynamic recrystallisation. This process takes place at a low homologous temperature of about 0.2 of the melting temperature (Mughrabi & Höppel, 2010). The structure is described as "bimodal". Dislocation patterns, characteristic for fatigue deformation of CG Cu, had developed in the coarser recrystallised grains. It is believed by Mughrabi & Höppel (2001) that this grain coarsening is closely related to the strain localisation. On the other hand, it is interesting that after fatigue at the plastic strain amplitude of 10^{-3} the grain coarsening was not observed.

Examination of the microstructure of failed specimens, which were used for the determination of the S-N curve of UFG Cu in Fig. 7, brought no evidence of structural changes, even for the highest stress amplitudes in LCF region. Fig. 24 is an example of a structure after stress symmetrical cycling at the stress amplitude of 255 MPa. The average grain size is 300 nm with the scatter usual for determination of the grain size in as ECAPed material. The corresponding cyclic plastic response during the test is shown in Figs. 17 and 18. Cyclic softening is a characteristic feature for the whole lifetime. This means that the cyclic softening is not directly related to the grain coarsening. This finding is in agreement with the observation by Agnew (1998) that the decrease in hardness of UFG Cu after fatigue does not scale with the cell size d_{cell} according to a well-known relationship between the saturation stress, $\sigma_{a,sat}$, and d_{cell} of the type $\sigma_{a,sat} \sim (d_{cell})^{-1/2}$. This suggests that the mechanism of softening is related to the decrease of defect density and changes of boundary misorientation and structure rather than to the gain size.

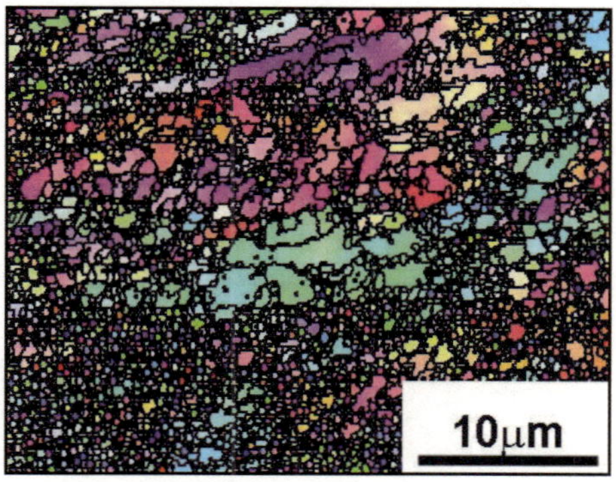

Fig. 27a. Microstructure as observed by EBSD, before fatigue

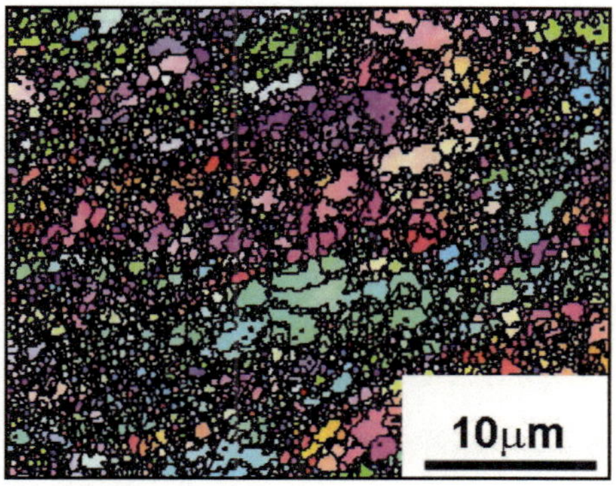

Fig. 27b. Microstructure as observed by EBSD, after fatigue

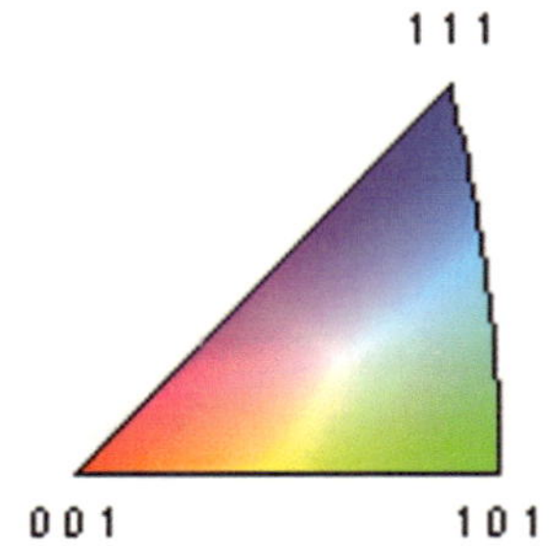

Fig. 27c. Colour code for inverse pole map

The highest number of cycles applied by the determination of the S-N curve, Fig. 7, was of the order of 10^{10} and was reached by ultrasonic loading at 20 kHz. Fig. 25 shows the TEM image of a structure of a specimen, which failed after 1.34 x 10^{10} cycles. Comparison with Fig. 2 implies no grain size changes after stress-controlled loading in gigacycle region. The characteristic cyclic stress-strain response in a very high-cycle region is continuous cyclic hardening; i.e., qualitatively different from that under cycling with high stress amplitudes.

Detailed analysis of many TEM micrographs tempts to believe that the fatigue loading with constant stress amplitude in the interval form 320 to 120 MPa, Fig. 7, does not result in the grain growth. The only observed structural change is a weak tendency to develop more "shaken down" dislocation structures (Kunz et al., 2006). An example is presented in Fig. 26.

The stability of the UFG structure of Cu on which was determined the S-N curve for symmetrical stress-controlled cycling, Fig. 7, and the S-N curve for tensile mean stress, Fig. 12, was examined by EBSD before loading and at failed specimens, far away from the fatigue crack. Similarly to TEM, this technique did not reveal any grain coarsening, even in the case of loading with the mean stress. EBSD, contrary to TEM on thin foils, enables observation of the development of microstructure on the same place. Fig. 27a shows the microstructure before fatigue loading. The microstructure is displayed in terms of a combination of an inverse pole figure map and a grain boundary network. Fig. 27b shows the same area on the failed specimen after fatigue loading with the stress amplitude 170 MPa and mean stress of 200 MPa. The colour key for identification of the grain orientation is given in Fig. 27c. From the comparison of Figs. 27a and b it is evident that the fatigue loading did not result in any grain coarsening, although pronounced cyclic softening was observed during fatigue. More likely the microstructure seems to be even finer after fatigue. Some larger grains decompose into more parts by development of new low angle boundaries. The detailed analysis of the area fraction occupied by grains of particular dimension before and after fatigue bears witness to this fact (Kunz et al., 2010).

In the case of UFG Cu significant differences in stability of structure were observed in dependence on the mode of fatigue testing. Generally, low stability of UFG structure was reported for plastic strain-controlled tests. The characteristic effect is formation of bimodal structure and shear banding (Höppel et al., 2009). Due to the obvious high sensitivity of fatigue behaviour of UFG Cu to internal and external parameters it is difficult to draw reliable conclusions from the comparison of literature data, which covers differently produced materials, different purity and different testing conditions. This is why on UFG Cu, on which the S-N curve in Fig. 7 was determined, the plastic strain controlled tests were conducted. The cyclic stress-strain responses corresponding to loading with ε_{ap} = 0.1 % and 0.05 % are shown in Fig. 19. An example of dislocation structures of material from the failed specimen loaded with ε_{ap} = 0.1 % is shown in Fig. 28. A well developed bi-modal microstructure consisting of areas with original fine-grained structure and large recrystallised grains with dislocation structure in their interior can be seen. This observation is in full agreement with results published by Mughrabi & Höppel (2001; 2010). The characteristic dislocation structure of a specimen loaded with the constant stress amplitude of 340 MPa is shown in Fig. 29. This structure does not exhibit any traces of bimodal structure, though the stress amplitude used is equal to the maximum value of the stress amplitude in the plastic strain amplitude-controlled test with ε_{ap} = 1 x 10^{-3}, Fig. 19. This means that the absolute value of the stress amplitude cannot be the reason for the substantially different stability of UFG structure under both types of tests. Also, the details of ECAP procedure are excluded. The tests were run on the same material. Also the frequency of loading in both tests was similar. The differences in the cumulative plastic strain amplitude in both tests were also not substantially different. The only difference between the two test modes, which can cause the different microstructure, seems to be the stress-strain response at the very beginning of the tests. There is relatively low plastic strain amplitude at the beginning of the stress-controlled test when compared to the strain-controlled test. It can be supposed that just the cycling with low strain amplitude at the beginning of the stress-controlled test can prevent the substantial changes of microstructure due to subsequent loading with increasing ε_{ap}. However, this idea is based on a small number of tests; further experimental study is necessary to support this opinion.

Fig. 28. Bi-modal dislocation structure after constant plastic strain amplitude loading

Fig. 29. Dislocation structure after constant stress amplitude loading, σ_a = 340 MPa

The research on stability of UFG Cu at higher temperatures has been aimed either at the investigation of the influence of elevated temperature during ECAP on the resulting microstructure or on the mechanical properties of UFG structure after post ECAP annealing. The best compromise between the tensile strength and ductility was achieved for Cu of 99.99 % purity prepared by route Bc after annealing at the temperature of 250 °C for 30 min. (Rabkin, 2005). Short annealing in the temperature range of 250 - 350 °C results in development of bi-modal structure consisting of large recrystallised grains embedded in fine-grained matrix. The thermal stability of UFG Cu after ECAP was found to be very low when compared to the cold rolled copper with the same total strain (Molodova, 2007). From the point of view of fatigue properties, the expectation that the bimodal grain size distribution should provide optimum fatigue performance is not justified (Mughrabi et al., 2006).

The up to now knowledge on the stability of UFG structure of Cu under cyclic loading is not sufficient to draw definite conclusions. On the other hand, it seems to be proven that the enhanced ductility and stabile microstructure are major facts that enhance the fatigue properties (Mughrabi et al., 2006). If the structural stability is low (due to internal material parameters or type of loading), the fatigue properties of UFG Cu are substantially reduced.

3.7 Fatigue crack initiation

Cyclic strain localisation resulting in fatigue crack initiation is an important stage of the fatigue process. It represents a substantial part of the fatigue life. Cyclic slip localisation results in a development of surface relief. Agnew & Weertman (1995) observed formation of slip bands on surface of fatigued specimens. Population of parallel cracks associated with extrusions develops during cycling. Their appearance resembles the surface relief well known from fatigue of CG Cu. Because the dimension of slip bands is substantially larger than the grain size of UFG structure and because they are oriented approximately 45°from the tension-compression axis, Agnew et al. (1999) denote them shear bands (SB). Since the early studies of the surface relief development, there are open questions concerning the nature and the mechanism of this phenomenon. The original belief that PSBs with the ladder like dislocation structure might be active in UFG Cu is dubious since the width of PSBs known from CG Cu is larger than the grain size of UFG structure. On the other hand, grain coarsening and development of bimodal structure was observed, particularly under plastic strain-controlled fatigue loading (Mughrabi & Höppel, 2010). Moreover, the dislocation patterns typical for fatigued CG Cu had developed in coarser grains. The relation of formation of SB and grain coarsening is not fully clarified up to now. There is an open question as to whether the process of shear banding is initiated by the local grain coarsening, which leads to the strain localisation, which destroys the original UFG structure, or the shear localisation takes place abruptly at first and the coarse structure is formed subsequently (Mughrabi & Höppel, 2001). Investigation of acoustic emission during cyclic deformation indicates that large-scale shear banding might be an important period of fatigue damage (Vinogradov et al., 2002).

The development of surface relief is a very common fatigue feature. An example of surface relief on fatigued UFG Cu is shown in Fig. 30. The observation was conducted on a specimen loaded in the HCF region. The material under investigation was the same as the material on which the S-N curve, Fig. 7, was determined; i.e., material which did not exhibit any grain coarsening under stress-controlled loading. Hence, the local coarsening of UFG structure is not the necessary prerequisite for formation of cyclic slip bands. Their length, see Fig. 30, substantially exceeds the grain size. Deep intrusions along the slip bands are visible. The extrusions rise high above the surface. The magistral fatigue crack develops by connection of suitably located intrusions. Observations of surface relief developed on specimens having fatigue life of the order of 10^{10} cycles show that the cyclic slip bands on the surface are very rare. An example of such bands is in Fig. 31. The cyclic slip bands produced by very high number of cycles are broad and make an impression of highly deformed local areas incorporating some neighbouring grains. The related intrusions are very short.

Investigations by Wu et al. (2003; 2004) on the relation of cyclic slip bands and the related microstructure beneath them did not reveal any grain coarsening. Fig. 32 shows an example of a focussed ion beam (FIB) micrograph of a cut perpendicular to the slip bands produced on the same copper on which the S-N curve in Fig. 7 Cu was determined. The surface relief (covered by protective Pt layer) and the underlying structure of material can be seen. No grain coarsening connected with the formation of fatigue slip bands can be stated. The appearance and size of the grains beneath the surface relief do not differ from those in other places. Numerous crack nuclei can be seen in this FIB micrograph. Some of them are directly connected with the surface roughness. In the material interior isolated cavities produced by cycling can be seen.

The rows of cavities below the surface slip bands seen in Fig. 32 can be considered to be nuclei of stage I cracks. Similar stage I cracks were observed by Weidner et al. (2010) in CG copper (grain size 60 microns) subjected to ultrasonic cycling not only in the surface grains, but also in the bulk grains. Similarity of both observations indicates that the mechanism of crack initiation in CG and UFG Cu in gigacycle region might be very similar. Substantial role in crack initiation will play point defects produced by dislocation interactions. They migrate along the grain boundaries and form row of cavities, which represent the crack nuclei.

The cyclic slip bands as observed in SEM by ion channelling contrast, Fig. 33, enable to correlate places of the cyclic strain localisation with the grain structure. This type of imaging visualises both the surface phenomena and the grain orientation. It can be seen that the cyclic slip bands lie in the zone where the grey contrast of neighbouring grains is low, which indicates that the disorientation between the grains is small. This zone can be called "zone of near-by oriented grains" (Kunz et al., 2011). The grains outside this zone obviously have a high mutual disorientation.

Fig. 30. Cyclic slip bands on the surface of UFG Cu loaded in HCF region

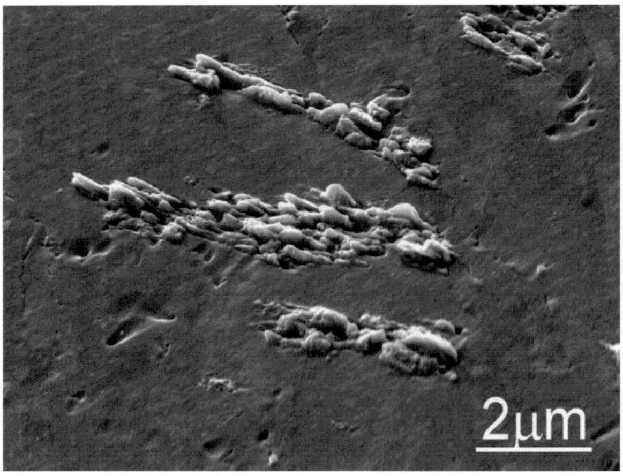

Fig. 31. Cyclic slip bands on the surface of UFG Cu loaded in gigacycle region

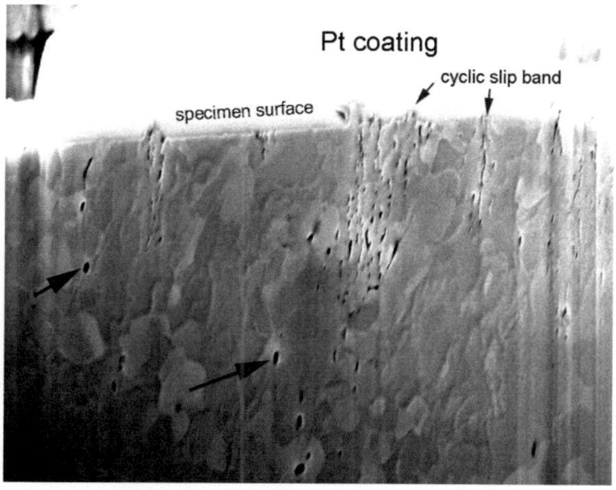

Fig. 32. FIB micrograph showing cut through a cyclic slip bands and grain structure

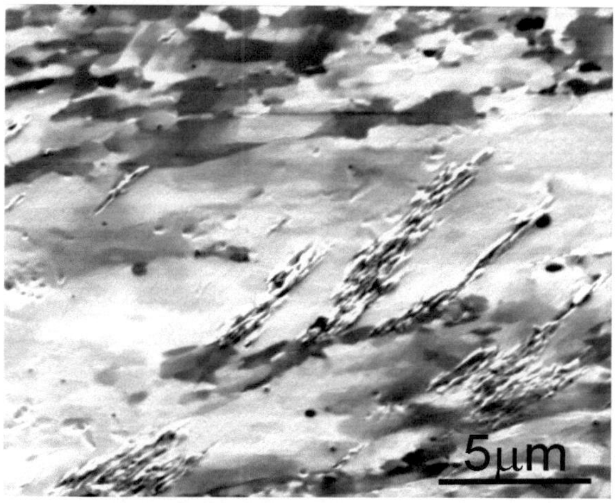

Fig. 33. SEM micrograph of surface slip bands using ion-induced secondary electron image

Based on the present-day state of knowledge the local grain coarsening is not a necessary condition for formation of cyclic slip bands and initiation of fatigue cracks. The slip bands, their shape and main features resemble the slip bands formed in CG Cu. Simultaneously, the grain coarsening is definitely an important effect taking place in UFG Cu under particular conditions. The role of the coarsened structure in the crack initiation process and the specific mechanism of initiation are not sufficiently understood. Contrary to the CG Cu, where the specific dislocation structures associated with cyclic slip bands are described thoroughly, there are no similar and conclusive observations on UFG Cu.

4. Conclusion

Severe plastic deformation can substantially improve the fatigue performance of Cu when cycled under stress-controlled conditions. The fatigue strength at 10^8 cycles can reach up to 150 MPa and 120 MPa for 10^{10} cycles, provided that the UFG microstructure remains stable. This depends both on material; i.e., the details of microstructure produced by SPD and also on the type of the cyclic loading. Under plastic strain-controlled tests, the UFG structure is more prone to grain coarsening and the fatigue life for the same plastic strain amplitude is substantially shorter than that of CG material. The plastic strain amplitude seems to be a unifying parameter for lifetime prediction. The fatigue cracks initiate at cyclic slip bands, which are observed under all types of loading and from the LCF to gigacycle region. With decreasing severity of cyclic loading their density decreases and their appearance slightly changes.

5. Acknowledgment

The Czech Science Foundation under contract 108/10/2001 financially supported this work. This support is gratefully acknowledged.

6. References

Agnew, S. R. & Weertman, J. R. (1998). Cyclic softening of ultrafine grain copper. *Mat. Sci. Eng. A*, Vol. 244, pp. 145-153, ISSN 0921-5093

Agnew, S. R., Vinogradov, A. Yu., Hashimoto, S. & Weertman, J. R. (1999). Overview of fatigue performance of Cu processed by severe plastic deformation. *J. Electronic Mater.*, Vol. 28, pp. 1038-1044, ISSN 0361-5235

Besterci, M., Kvačkaj, T., Kováč, L. & Sűlleiová, K. (2006). Nanostructures and mechanical properties developed in copper by severe plastic deformations. *Kovove Mater.*, Vol. 44, pp. 101-106, ISSN 0023-432X

Goto, M., Han, S. Z., Yakushiji, T., Kim, S. S. & Lim, C. Y. (2008). Fatigue strength and formation behavior of surface damage in ultrafine grained copper with different non-equilibrium microstructures. *Int. J. Fatigue*, Vol. 30, pp. 1333-1344, ISSN 0142-1123

Goto, M., Han, S. Z., Kim, S. S., Ando, Y. & Kwagoishi, N. (2009). Growth mechanism of a small surface crack of ultrafine-grained copper in a high-cycle fatigue regime. *Scripta Mat.*, Vol. 60,pp. 729-732, ISSN 1359-6462

Han, S. Z., Goto, M., Lim, Ch., Kim, S. H. & Kim, S. (2007). Fatigue behavior of nano-grained copper prepared by ECAP. *J. of Alloys and Comp.*, Vols. 434-435, pp. 304-306, ISSN 0925-8388

Han, S. Z., Goto, M., Lim, Ch., Kim, Ch. J. & Kim, S. (2009). Fatigue damage generation in ECAPed oxygen free copper. *J. of Alloys and Comp.*, Vol. 483, pp. 159-161, ISSN 0925-8388

Hashimoto, S., Kaneko, Y., Kitagawa, K., Vinogradov, A. & Valiev, R. (1999). On the cyclic behaviour of ultra-fine grained copper produced by equi-channel angular pressing. *Mat. Sci. Forum*, Vol. 312-314. pp. 593-598, ISSN 0255-5476

Höppel, H. W., Brunnbauer, M., Mughrabi, H., Valiev, R. Z. & Zhilyaev, A. P. (2000). Cyclic deformation behaviour of ultrafine grain size copper produced by equal channel angular pressing. *Materials Week 2000*, Munich 2001. Available from: <http://www.materialsweek.org/proceedings>

Höppel, H. W., Zhou, Z. M., Mughrabi, H., &Valiev, R. Z. (2002). Microstructural study of the parameters governing coarsening and cyclic softening in fatigued ultrafine-grained copper. *Phil. Mag.*, Vol. 82, pp. 1781-1794, ISSN 0141-8610

Höppel, H. W. & Valiev, R. Z. (2002). On the possibilities to enhance the fatigue properties of ultrafine-grained metals. *Z. Metallkd.*, Vol. 93, pp. 641-648, ISSN 0044-3093

Höppel, H. W., Kautz, M., Xu, C., Murashkin, M., Langdon, T. G., Valiev, R. Z. & Mughrabi, H. (2006). An overview: Fatigue behaviour of ultrafine-grained metals and alloys. *Int. J. Fatigue*, Vol. 28, pp. 1001-1010, ISSN 0142-1123

Höppel, H. W., Mughrabi, H. & Vinogradov, A. (2009). Fatigue properties of bulk nanostructured materials. In: *Bulk Nanostructured materials*, Zehetbauer, M. et al. (Eds.), Wiley-VCH Verlag, Weinheim, pp. 481-500. ISBN 978-3-527-31524-6

Klesnil, M. & Lukáš, P. (1992). *Fatigue of metallic materials.* ISBN 0-444-98723-1, Academia/Elsevier, Prague, Czech Republic

Kunz, L., Lukáš, P. & Svoboda, M. (2006). Fatigue strength, microstructure stability and strain localization in ultrafine-grained copper. *Mat. Sci. Eng. A*, Vol. 424, pp. 97-104, ISSN 0921-5093

Kunz, L., Lukáš, P., Pantělejev, L. & Man, O. (2010). Stability of microstructure of ultrafine-grained copper under fatigue and thermal loading. *Strain.* doi: 10.1111/j.1475-1305.2009.00710.x, online ISSN 1475-1305

Kunz, L., Lukáš, P., Pantělejev, L. & Man, O. (2011). Stability of ultrafine-grained structure of copper under fatigue loading. *Procedia Engineering*, Vol. 10, pp. 201-206. ISSN: 1877-7058

Kuokkala, V. T. & Kettunen, P. (1985). Fatigue of polycrystalline copper at constant and variable plastic strain amplitudes. *Fat. Fract. Eng. Mater. and Struct.*, Vol. 8, pp. 277-285, ISSN 8756-758X

Kwan, Ch. C. F. & Wang, Z. (2011). On the cyclic deformation response and microstructural mechanisms of ECAPed and ARBed copper - an overview. *Mat. Sci. Forum*, Vol. 683, pp. 55-68 ISSN 1662-9752

Langdon, T. G., Furukawa, M., Nemoto, M. & Horita, Z. (2000). Using equal-channel angular pressing for refining grain size. *JOM*, April, pp. 30-33, ISSN 1047-4838

Lukáš, P. & Klesnil, M. (1973). Cyclic stress-strain response and fatigue life of metals in low amplitude region. *Mat. Sci. Eng.*, Vol. 11, pp. 345-356, ISSN 0025-5416

Lukáš, P. & Kunz, L. (1985). Is there a plateau in the cyclic stress-strain curves of polycrystalline copper? *Mat. Sci. Eng.*, Vol. 74, L1 - L5, ISSN 0025-5416

Lukáš, P. & Kunz, L. (1987). Effect of grain size on the high cycle fatigue behaviour of polycrystalline copper. *Mat. Sci. Eng.*, Vol. 85, pp. 67-75, ISSN 0023-5416

Lukáš, P. & Kunz, L. (1988). Effect of low temperatures on the cyclic stress-strain response and high cycle fatigue life of polycrystalline copper. *Mat. Sci. Eng. A*, Vol. 103. pp. 233-239, ISSN 0921-5093

Lukáš, P., Kunz, L. & Svoboda, M. (2007). Effect of low temperature on fatigue life and cyclic stress-strain response of ultrafine-grained copper. *Met. Mat. Trans. A*, Vol. 38A, pp. 1910-1915. ISSN: 1073-5623

Lukáš, P., Kunz, L., Svoboda, M. (2008). Fatigue of ultrafine grained copper. In: *Proc. of the 6th int. conf. on low cycle fatigue (LCF6)*. Portella, P. D. et al. (Eds.), Berlin, DVM, pp. 295-306.

Lukáš, P., Kunz, L., Svoboda, M., Buksa, M. & Wang, Q. (2008). Mechanisms of cyclic plastic deformation in ultrafine-grain copper produced by severe plastic deformation. In: *Plasticity, Failure and Fatigue in Structural Materials - from Macro to Nano. Proc. of the Hael Mughrabi Honorary Symposium.* Hsia et al. (Eds.), TMS, pp. 161- 166, ISBN 978-0-87339-714-8

Lukáš, P., Kunz, L. & Svoboda, M. (2009). Fatigue mechanisms in ultrafine-grained copper. *Kovove Mater.*, Vol. 47, pp. 1-9, ISSN 0023-432X

Mingler, B., Karnthaler, H. P., Zehetbauer, M. & Valiev, R. Z. (2001). TEM investigation of multidirectionally deformed copper. *Mat. Sci. Eng. A*, Vol. 319-321, pp. 242-245, ISSN 0921-5093

Mishra, A., Richard, V., Grégori, F., Asaro, R. J. & Meyers, M. A. (2005). Microstructural evolution in copper processed by severe plastic deformation. *Mat. Sci. Eng. A*, Vol. 410-411, pp. 290-298, ISSN 0921-5093

Molodova, X., Gottstein, G. & Hellmig, R. J. (2007). On the thermal stability of ECAP deformed fcc metals. *Mater. Sci. Forum*, Vols. 558-559, pp.259-264

Mughrabi, H. (1978). The cyclic hardening and saturation behaviour of copper single crystals. *Mat. Sci. Eng.*, Vol. 33, pp. 207-223, ISSN 0025-5416

Mughrabi, H. & Höppel, H. W. (2001). Cyclic deformation and fatigue properties of ultrafine grain size materials: current status and some criteria for improvement of the fatigue resistance. In: *Mat. Res. Soc. Symp. Proc.* Farkas, D. et al. (Eds.), Mater. Res. Soc. Warrendale Penn. Vol. 634, pp. B2.1.1-B2.1.12

Mughrabi, H., Höppel, H. W. & Kautz, M. (2004). Fatigue and microstructure of ultrafine-grained metals produced by severe plastic deformation. *Scripta Mater.*, Vol. 51, pp. 807-812, ISSN 1359-6462

Mughrabi, H., Höppel, H. W. & Kautz, M. (2006). Microstructural mechanisms governing the fatigue performance of ultrafine-grained metals and alloys. *In: Ultrafine grained materials IV*, Zhu, Y. T. et al. (Eds.), TMS, pp. 47-54

Mughrabi, H. & Höppel, H. W. (2010). Cyclic deformation and fatigue properties of very fine-grained metals and alloys. *Int. J. Fatigue*, Vol. 32, pp. 1413-1427, ISSN 0142-1123

Müllner, H., Weiss, B., Stickler, R., Lukáš, P. & Kunz, L. (1984). The effect of grain size and test frequency on the fatigue behavior of polycrystalline Cu. *Fatigue 84. Proc. of the 2nd Int. Conf. on Fatigue Thresholds*, p. 479.

Murphy, M. C. (1981). The engineering fatigue properties of wrought copper. *Fatigue of Engng. Mater and Struct.*, Vol. 4, pp. 199-234, ISSN 0160-4112

Polák, J. & Klesnil, M. (1984). Cyclic stress-strain response and dislocation structures in polycrystalline copper. *Mat. Sci. Eng.*, Vol. 63, pp. 189-196, ISSN 0025-5416

Rabkin, E., Gutman, I., Kazakevich,.M., Buchman, E. & Gorni, D. (2005). Correlation between the nanomechaical properties and microstructure of ultrafine-grained copper produced by equal channel angular pressing. *Mat. Sci. Eng. A*, Vol. 396, pp. 11-21. ISSN 0921-5093

Saada, G. (2005). Hall-Petch revisited. *Mat. Sci. Eng. A*, Vols. 400-401, pp. 146-149, ISSN 0921-5093

Segal, V. M. (1995). Materials processing by simple shear. *Mat. Sci. Eng. A*, Vol. 197, pp. 157-164, ISSN 0921-5093

Thompson, A. W. & Backofen, W. A. (1971). The effect of grain size on fatigue. *Acta Met.*, Vol. 19, June, pp. 597-606, ISSN 0001-6160

Valiev, R. Z., Islamgaliev, R. K. & Alexandrov, I. V. (2000). Bulk nanostructured materials from severe plastic deformation. *Prog. Mater. Sci.*, Vol. 45, pp. 103-189, ISSN 0079-6425

Valiev, R. Z. & Langdon, T. G. (2006). Principles of equal-channel angular pressing as a processing tool for grain refinement. *Prog. Mater. Sci.*, Vol. 51, pp. 881-981, ISSN 0079-6425

Vinogradov, A., Kaneko, Y., Kitagawa, K., Hashimoto, S., Stolyarov, V. & Valiev, R. (1997). Cyclic response of ultrafine-grained copper at constant plastic strain amplitude. *Scripta Mater.*, Vol. 36, pp. 1345-1351, ISSN 1359-6462

Vinogradov, A. & Hashimoto, S. (2001). Multiscale phenomena in fatigue of ultra-fine grain materials - an overview. *Mater. Trans.*, Vol. 42, pp. 74-84, ISSN 0916-1821

Vinogradov, A., Hashimoto, S., Patlan, V. & Kitagawa, K. (2001). Atomic force microscopic study on surface morphology of ultra-fine grained materials after tensile testing. *Mat. Sci. Eng. A*, Vols. 319-321, pp. 862-866, ISSN 0921-5093

Vinogradov, A. & Hashimoto, S. (2002), Fatigue of severely deformed metals. *In: Nanomaterials by severe plastic deformation. Proc. of the conf. Nanomaterials by severe plastic deformation.* DGM, Wiley-VCH Verlag, pp. 663-676

Vinogradov, A., Patlan, V., Hashimoto, S. & Kitagawa K. (2002). Acoustic emission during cyclic deformation of ultrafine-grained copper processed by severe plastic deformation. *Phil. Mag.*, Vol. 82, pp.317-335

Wang, J.T. et al. (Eds.). (2011). *Nanomaterials by severe plastic deformation: NanoSPD5.* TTP Publications LTD, Switzerland. ISSN 0255-5476

Wang, Z. & Laird, C. (1988). Cyclic stress-strain response of polycrystalline copper under fatigue conditions producing enhanced strain localization. *Mat. Sci. Eng.*, Vol. 100, pp. 57-68, ISSN 0025-5416

Weidner, A., Amberger, D., Pyczak, F., Schönbauer, B., Stanzl-Tschegg, S. & Mughrabi, H. (2010). Fatigue damage in copper polycrystals subjected to ultrahigh-cycle fatigue below the PSB threshold. *Int. J. Fatigue*, Vol. 32, pp. 872-878

Wilkinson, A. J. & Hirch, P. B. (1997). Electron diffraction based techniques in scanning electron microscopy of bulk materials. *Micron*, Vol. 28, pp. 279-308, ISSN 0968-4328

Wu, S. D., Wang, Z. G., Jiang, C. B., Li, G. Y., Alexandrov, I. V. & Valiev, R. Z. (2003). The formation of PSB-like shear bands in cyclically deformed ultrafine grained copper processed by ECAP. *Scripta Mater.*, Vol. 48, pp. 1605-1609, ISSN 1359-6462

Wu, S. D., Wang, Z. G., Jiang, Li, G. Y., Alexandrov, I. V. & Valiev, R. Z. (2004). Shear bands in cyclically deformed ultrafine grained copper processed by ECAP. *Mat. Sci. Eng. A*, Vols. 387-389, pp. 560-564.

Xu, Ch., Wang, Q., Zheng, M., Li, J., Huang, M., Jia, Q., Zhu, J., Kunz, L. & Buksa, M. (2008). Fatigue behavior and damage characteristic of ultra-fine grain low-purity copper processed by equal-channel angular pressing (ECAP). *Mat. Sci. Eng. A*, Vol. 475, pp. 249-256, ISSN 0921-5093

Zhu, Y. T. & Langdon, T. G. (2004). The fundamentals of nanostructured materials processed by severe plastic deformation. JOM, October, pp. 58-63, ISSN 1047-4838

6

Fatigue Crack Resistance of Ultrafine-Grained Copper Structures

Luca Collini
Department of Industrial Engineering, University of Parma
Italy

1. Introduction

In the last 25 years, we have witnessed to an increasing interest in developing and characterising the so-called micro- and nano-scale classes of structural materials. Micro- and nano-materials, which have arrangements, grain structures and sub-structures of less than one micron, present mechanical properties better than those belonging to the original materials (Valiev, 1997; Valiev et al., 2000). In general, since the grain size ranges between 0.1 and 1 μm, such materials are designated as ultrafine-grained (UFG) materials.

Nowadays, UFG materials have many technological applications. Due to their outstanding mechanical properties they can be successfully employed in aircraft construction as well as for high performance tools. The highly localised shear in the grain refinement process enhances their behaviour, when used as a simple plastic joint and when employed as self-sharpening tools. There are two main methods to produce UFG metals. The so-called "bottom-up" approach of synthesis, by which atoms, molecules and nanoparticles participate in order to build blocks for the creation of complex structures. Alternatively, there is the "top-down" process in which a hard solid-state elaboration of materials occur. In this approach, coarse-grained materials are refined into nanostructured materials through heavy straining or shock loading. In particular, as the market asks more and more for UFG, researchers must answer the big questions related to the complete identification of their mechanical behaviour.

Presently, the most studied nanomaterials are titanium, aluminium and copper alloys, which are also those mainly considered in this work. The ultrafine-grained Copper (UFG-Cu) structure is obtained from coarse-grained Copper (CG-Cu), from which it maintains the FCC crystallographic system, through different techniques. Among these, the most popular are electrodeposition, mechanical alloying, and severe plastic deformation (SPD) techniques. This last class of technique belongs to the Equal Channel Angular Pressing (ECAP) method, which consists of the cold refinement of the grain structure, obtained by the repeated passage of a material billet through an angular die (Valiev & Langdon, 2006). One can observe that, as it can be seen in Fig. 1, the UFG-Cu obtained from an increasing number of ECAP passages tendentially presents a structure made up of equiaxed grains, but of a medium size and strongly reduced with respect to the original CG-Cu. In fact, the resulting structure is composed of grains of different sizes, and it is important to note that the presence of larger grains still provides ductility to the material (Wei & Chen, 2006).

To give an appropriate and complete description of the advanced characteristics of this class of materials, in line with the technological and process research, much work has still to be done in the characterisation of the mechanical properties of UFG structures, especially in those concerning fatigue and crack propagation resistance. Furthermore, fracture mechanisms and damage evolution have yet to be fully understood. Some relevant works describing studies on nanomaterials are those presented in Meyers et al. (2006) and Valiev (1997), while others focused on the investigation of the fatigue and the fatigue crack resistance of UFG-Cu are Lugo et al. (2008), Höppel et al. (2006), Cavaliere (2009), Lukáš et al. (2009), Vinogradov (2007) and Collini (2010a, 2010b).

First of all, one cannot disregard the size effect derived from the intense refining of the grains in the material: in fact, in reducing the typical sizes of the structure below one micron, new deformation mechanisms will be taking place and this leads to behaviours that are different from those characteristic of the CG structures. In particular, the size effect is due to the formation of new types of dislocation such as the Geometrically Necessary (GNDs) type, and it causes a shifting from the Hall-Petch relation, and furthermore plastic deformation does not appear uniform for such a small structure. More relevant, is that the constitution and the interaction between the grain interior regions and the grain boundary regions assumes an increasing importance as the typical sizes of the structures decrease, because the grain boundary gradually becomes thicker and represents a stiff wall around the inside of the grain, in which even changes to the mechanisms of the dislocations pile-up (Kozlov et al., 2004).

The aim of the hardening process in a metal, i.e. of the increasing of the local plastic deformation energy, is to increase the amount of energy that forces a dislocation to move inside a single grain (pile-up mechanism), and from one grain to another crossing a grain boundary. In fact, the higher the stress needed to invoke the dislocation movement, the higher the yield strength of the metal. This is why the dislocation hardening produced by grain refinement brings a considerable enhancement in static and fatigue strength, and of the hardness of metals (Thomson & Backofen, 1971; Meyers et al., 2006; Mughrabi et al., 2004).

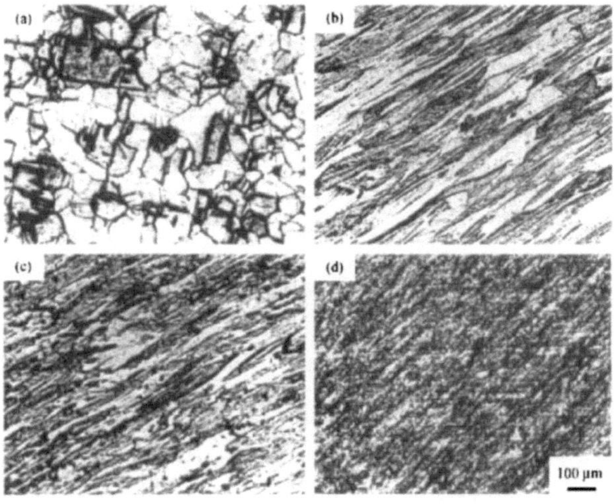

Fig. 1. Microcrystalline structure of copper: (a) before and (b-d) after 8, 10 and 12 ECAP passages (Wei & Chen, 2006).

The effect of the grain size on the cyclic plasticity and fatigue life of metals has been the focus of many investigations on steel, copper, nickel, titanium and magnesium based alloys. In general, two major conclusions based on these studies have been drawn: (i) the fatigue limit of pure f.c.c. metals with relatively high stacking fault energy and wavy slip behaviour is not affected by the grain size; and (ii) the fatigue strength of materials exhibiting planar slip, increases with decreasing grain size and follows the Hall-Petch relationship in the same way as the yield stress in conventional polycrystalline metals (Vinogradov, 2007). In particular, the studies concerned with UFG copper have shown the following aspects: 1) the UFG copper exhibits higher fatigue strength than the coarse-grained (CG) counterpart when the cycling is stress-controlled (Kunz et al., 2006), and presents lower fatigue strength during strain-controlled testing (Höppel et al., 2006); 2) a low purity UFG copper alloy shows higher resistance in its overall fatigue life (Xu et al., 2008); 3) the level of purity affects the fatigue behaviour (Lukáš et al., 2009). It has been found that the ECAP route also affects fatigue strength. Both the effects (purity and route) are more pronounced at low stress amplitudes.

A large number of studies have been conducted on the static and fatigue properties of UFG copper, but up until now very little data on the fatigue crack growth (FCG) behaviour of this material are available. This is mainly due to the difficulty in obtaining quite large bulk volumes of an ECAPed material to be machined in standard specimens for crack propagation tests. However, knowledge of FCG behaviour is of crucial importance for most engineering applications, and it is necessary for a comprehensive understanding of the fatigue properties.

Experimental fatigue crack propagation curves for the UFG copper can be found in the literature (Vinogradov, 2007; Cavaliere, 2009; Horky et al. 2011). These works show that UFG materials exhibit the same crack propagation behaviour as polycrystals, i.e. a threshold regime, an intermediate stable growth regime well described by the Paris-Erdogan law, and an instable regime at high crack growth rates. In UFG Ni, the growth rate of a defect in the threshold regime is higher in a UFG alloy than in that of the polycrystalline reference material (Hanlon et al., 2005). This behaviour has been ascribed to the absence of a tightening mechanism in the UFG state, e.g. roughness of the crack path, due to its peculiar microstructure. Indeed, in UFG microstructures the crack path usually appears as straight and smooth, providing faster growth rates under limited crack tip plasticity. A prevalence of intergranular fracture modes during FCG was experimentally observed, explaining the nearly straight crack path in a uniform UFG structure. However, direct in-situ observation of the initiation and growth of a small, semi-elliptical surface cracks in a UFG copper structure has shown a transition of the propagation mechanism after about 0.1 mm of crack length: as the schematic in Fig. 2 illustrates. From intergranular and straight, the crack path becomes tortuous with a decrease in the FCG rate when the cyclic plastic zone (CPZ) ahead of the crack tip involves quite a large number of grains (Goto et al., 2009). This change in the propagation mechanism is due to the interrelationship between the grains' structure and the crack. Another work on high-purity UFG copper prepared by high pressure torsion (HPT) shows that the resistance to crack initiation is increased by the grain refinement, but that the stability of the microstructure during cycling is of utmost importance (Horky et al. 2011).

However, a deeper understanding of the resistance of ultrafine-grained structures is necessary, in particular about the propagation of long, well-developed cracks, and open questions remain about the role of the reverse plastic zone at the crack tip, the shear bands formation mechanism and their interaction with the specific, small-scale granular structure, and possible toughening mechanisms.

In this chapter, the fatigue crack growth resistance of a copper alloy in the UFG state with a commercial purity level will be presented and discussed, in the light of recent findings. Results of the laboratory activity are juxtaposed with data from technical literature referring to coarse and ultrafine copper with different purity levels. A discussion on the propagation mechanisms will be conducted with the support of simplified closure and toughening models derived from the SEM analysis of crack paths.

2. Effect of grain size in crack propagation resistance

A decrease in grain size of metals and alloys generally results in an increase in strength. Improved ultimate tensile strength and yield strength does not necessarily mean that the fine-grained materials also exhibit better properties with respect to their resistance against damage by fatigue crack growth.

Propagation of long cracks in conventionally grained Cu can be well described by linear facture mechanics. There is a good correlation between the crack propagation rate, *da/dN*, (*a* is the crack length and *N* number of cycles) and the stress intensity factor *K*. The experimentally determined crack propagation curve for stress symmetrical loading is shown in Fig. 3. The crack rate was determined on sheet centre-cracked tension specimens manufactured from Cu of 99.99 % purity. The cycling was conducted under load-controlled conditions. It is evident that the grain size influences the crack rate. Cracks in the fine-grained Cu with a grain size of 70 μm propagate faster than in the coarse-grained Cu with a grain size of 1.2 mm (Lukáš & Kunz, 1986). The experimental results in Fig. 3 can be well approximated by the equation:

$$da / dN = A(K_a^m - K_{ath}^m) \tag{1}$$

where A = 1.1 x 10^{-10} (mm/cycle)(MPa√m)$^{-m}$ and m = 7.0 for fine-grained Cu and A = 5.0 x 10^{-11} (mm/cycle)(MPa√m)$^{-m}$ and m = 7.1 for coarse-grained Cu. The threshold value of SIF, below which the long cracks do not propagate, is K_{ath} = 2.1 and 2.7 MPa√m for fine-grained and coarse-grained Cu, respectively. From this data it is evident that from the point of view of crack propagation, the fine-grained Cu is worse than the coarse-grained.

There are a plenty of models for the propagation of long cracks. They can be divided into three basic groups: 1) models assuming that the plastic deformation in the plastic zone is the determining factor in crack growth, 2) models based on damage in front of the crack tip and 3) models based on energy considerations. The knowledge obtained from the investigation into crack propagation in Cu clearly favours those models based on the crucial role of the cyclic plastic strain at the crack tip.

In Cu of conventional grain size, a fatigue crack tip is surrounded by a dislocation cell structure in a plastic zone. An example of a dislocation structure adjacent to the long fatigue crack propagating in the Paris law region (a linear part of the *da/dN* vs. K_a plot in log-log coordinates) can be seen in Fig. 4. The fatigue crack was propagating in a non-crystallographic manner, macroscopically perpendicular to the principal stress. The foils for transmission electron microscopy (TEM) were prepared in the following manner. The fatigue crack surface was electrodeposited. The foils for TEM were prepared from thin slices cut perpendicularly to the macroscopic fracture surface and parallel to the crack propagation direction. Because the crack length corresponding to the location of the foils and cyclic loads were known, it was possible to correlate the observed structure with K_a.

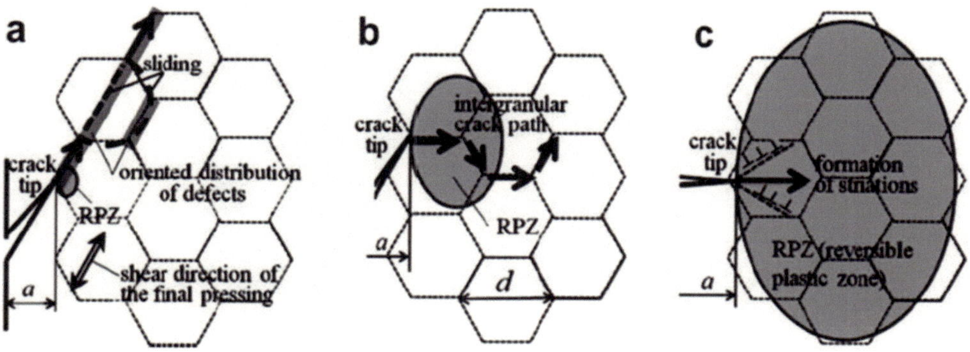

Fig. 2. Schematic of the FCG mechanism of a small surface crack in UFG Cu and of the relationship between the cyclic plastic zone (CPZ) at the crack tip and grain size: (a) when CPZ is smaller than the grain size, the crack propagates in a mechanism conforming to the local area (grain or grain boundary), e.g. along the final shear pressing direction; (b) when CPZ is 1-2 times the grain size, the crack grows along GBs where an incompatibility in deformation in adjacent grains is concentrated, showing an intergranular crack path; (c) when the CPZ is more than 3-4 times the grain size, the crack propagates due to the striation formation mechanism, associated with crack tip retardation and blunting. From Goto et al. (2009).

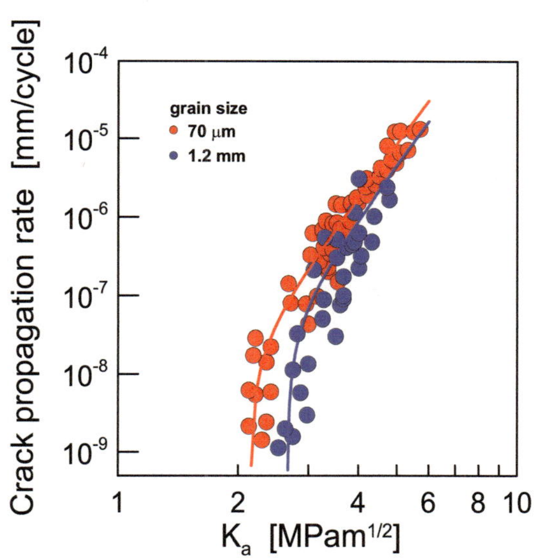

Fig. 3. Comparison of crack propagation curves for fine-grained and coarse-grained Cu.

On the left hand side of Fig. 4 the electrodeposit can be seen. On the right hand side is the well-developed cell structure, which was produced by the cyclic plastic deformation in the plastic zone. The original fatigue fracture surface can be seen in-between. Analysis of a number of TEM foils indicates that the cell size d is inversely proportional to K_a (Lukáš & Kunz, 1986).

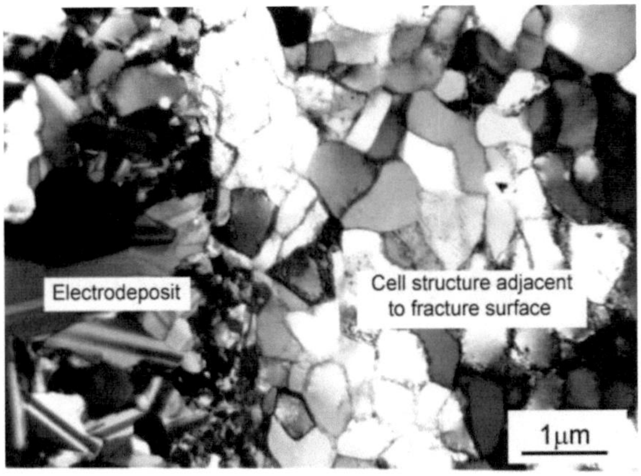

Fig. 4. Dislocation cell structure adjacent to fracture surface in conventionally grained Cu.

For the lowest crack growth rates of the order of 10^{-9} mm/cycle in the threshold region when K_a approaches K_{ath} the crack propagation mechanism in Cu changes. The crack propagates in a "zig-zag" manner and the fracture surface exhibits crystallographic features. The region of plastic zone with the cell structure is very small. TEM observations, an example is shown in Fig. 5, indicate that the crack propagates along the persistent slip bands (PSB) with a characteristic ladder-like structure. The fracture surface is straightforward and runs along the ladder. This means that at very low rates the propagating crack does not necessarily change the vein dislocation structure characteristic of the material in which the crack propagates.

Based on the observation of the dislocation structures near the crack tip in Cu, the following considerations can be made. Fig. 6 schematically illustrates the situation at the crack tip. At high crack growth rates and small grain sizes, the "fracture mechanical" plastic zone consisting of the cell structure is large compared to the length of the PSBs, which develops before the propagating crack tip. The plastic strain amplitude at the crack tip controls the crack growth process. However, for low crack growth rates and coarse-grained Cu the ratio of the length of PSBs terminating at the grain boundaries to the cell zone size increases. Some observations show that the cell zone can vanish completely. The decisive majority of the cyclic plastic strain is concentrated in the PSBs. This type of plastic zone is different from the fracture mechanical zone, which develops at loading with a high K_a. It is obvious that this effect and the effect of the grain size can substantially influence the crack growth rate.

UFG Cu produced by ECAP is typically has a grain size (cell size) of about 300 nm. This is comparable to the smallest cells observed at the tip of the fatigue crack propagating at the conditions characterized by $K_a \sim 10$ MPa√m (Lukáš et al., 1985). From this point of view the crack propagation resistance of UFG and CG copper in this region should be similar. Strong differences, however, can be expected in the threshold region. Here the stability of the UFG structure would play the decisive role. Horky et al. (2011) studied the crack propagation in UFG Cu prepared by high pressure torsion. They observed very expressive grain coarsening in the vicinity of the fatigue crack in Cu of high purity. Retardation of the crack growth was reported.

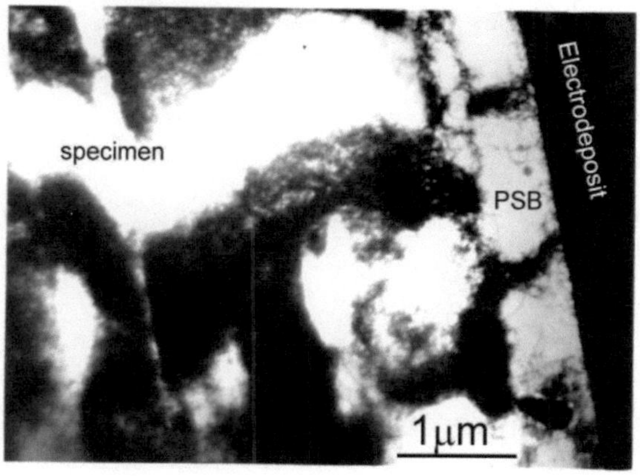

Fig. 5. Dislocation structure adjacent to the fracture surface in conventionally grained Cu created by extremely low crack growth.

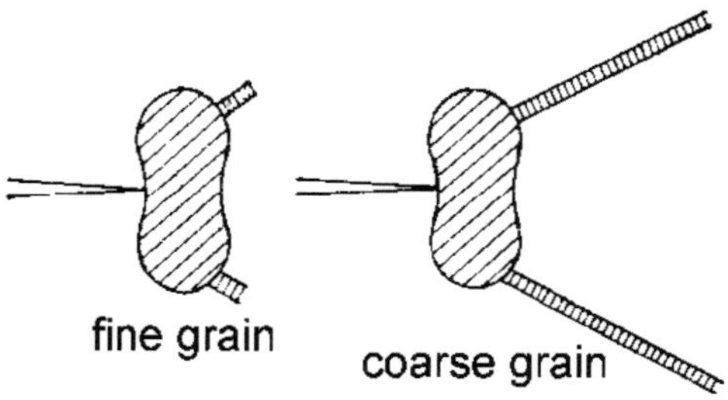

Fig. 6. Schematic representation of plastic zone in fine-grained and coarse-grained Cu.

3. Experimental methodology

3.1 Material

The material used for this study is a copper alloy, subjected to 8 passes ECAP through route Bc, i.e. with a billet rotation in the same sense rotated by 90° between each pass. The copper has been processed in the laboratory of Prof. R.Z. Valiev at the Ufa State Aviation Technical University (Russia), starting from cylindrical samples of 20 mm in diameter and 120 mm in length. Cylindrical samples of 16 mm in diameter and 100 mm in length were machined from the billets.

The microstructure obtained is shown in the micrograph of Fig. 7: the result is fine and uniform, with grain sizes ranging from 100 to 800 nm and an average size of 300 nm. The orientation map, visible in Fig. 7 on the right, shows low angle grain boundaries. The chemical composition, reported in Tab. 1, indicates that the purity, 99.90%, is of a commercial level.

The static and fatigue properties are summarised in Tab. 2, in which the properties of CG wrought copper alloy (grain size 30 μm) are also reported (Murphy, 1981). As can be seen, the ECAP process brings evident advantages to the mechanical properties: the yield strength σ_y is 4-times that of the CG, and the fatigue limit at 10^8 cycles σ_f is twice the original.

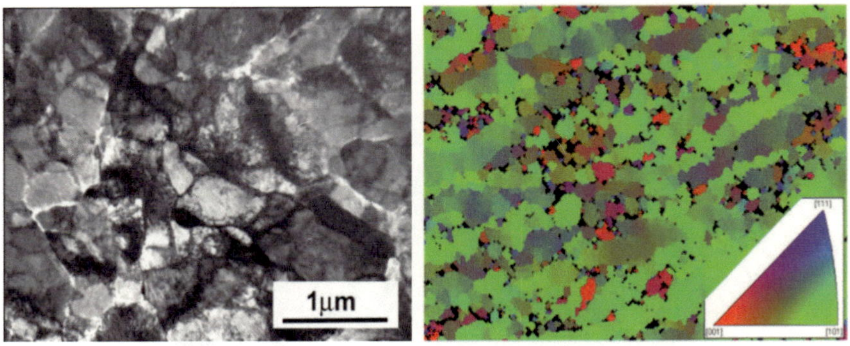

Fig. 7. TEM micrograph and grain orientation map of the ECAPed UFG copper.

Bi	Sb	As	Fe	Ni	Pb	Sn	S	O	Zn	Ag	Cu
0.001	0.002	0.002	0.005	0.002	0.005	0.002	0.004	0.05	0.004	0.003	balance

Table 1. Chemical composition (% in wt) of the UFG copper.

Structure	Grain size d_G (μm)	σ_y (MPa)	σ_u (MPa)	A%	σ_f at 10^8 cycles (MPa)
CG	30	95±5	195±5	41.5	77
UFG	0.300±0.015	375±4	387±5	18.5	168

Table 2. Mechanical properties of CG and UFG polycrystalline copper.

3.2 Procedure and data elaboration

The FCG tests were conducted in laboratory on Disk Shaped CT specimens, in air and at room temperature. A MTS 810 servo-hydraulic machine working at a frequency of 10 Hz has been used for the test. A view of the test apparatus is shown in Fig. 8.

Specimens were machined in discs of 7 mm in thickness from bars of 16 mm in diameter. The DSCT specimen is shown in Fig. 9. Due to its reduced dimensions, it has been necessary to use a Back Face Strain Measurement (BFSG) technique to monitor the crack length on the specimen, by using a very small strain gage glued to its back. In conjunction with a finite element calibration, the BFSG technique allows the calculation of the mode I Stress Intensity Factor (SIF) as a function of a/W ratio by Eq. (2):

$$K_I = \frac{P\left(2+\frac{a}{W}\right)}{BW^{1/2}\left(1-\frac{a}{W}\right)^{3/2}} \sum_{i=0}^{4} C_i \left(\frac{a}{W}\right)^i \tag{2}$$

where P is the applied load, B and W are the characteristic specimen dimensions, and C_i is the constant relative to the DSCT geometry. The standard procedure for FCG calculation is adopted in order to realize K-increasing and K-decreasing (load-shedding) tests. Experimental tests are conducted at a load ratio $R = K_{min}/K_{max}$ equal to 0.1, 0.3, 0.5 and 0.7. The mode I crack propagation resistance is investigated both in the stable growth regime (stage II), and in the threshold regime (stage I).

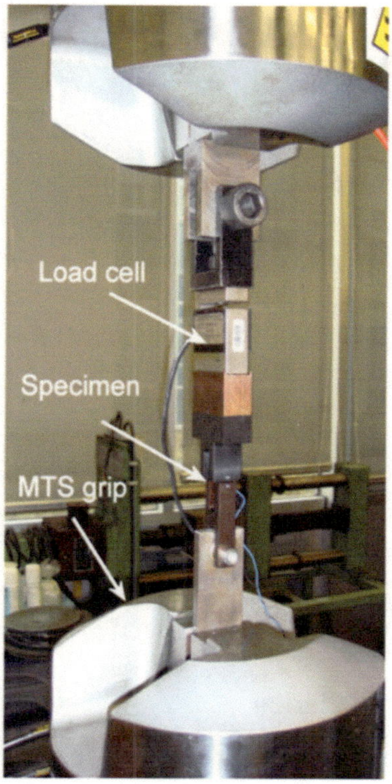

Fig. 8. FCG test apparatus.

Acquisition of each data point is automatically made every 0.1 mm of crack propagation. In order to facilitate the crack initiation and mode I crack propagation, an initial fatigue pre-cracking of 0.8 mm at load ratio 0.1 is conducted for all specimens. An estimation of ΔK_{th} at the conventional growth rate of 10^{-7} mm/cycle has been made following the procedure of regularisation. It means that data points of crack a as a function of applied ΔK, and of ΔK itself as a function of the number of cycles N, have been interpolated by exponential curves, adopting opportune values for parameters $\kappa_{1,2}$ and $\beta_{1,2}$, as indicated in Eqs. (3,4):

$$a(N) = a_0 + \kappa_1\left(1 - e^{\kappa_2 N}\right) \tag{3}$$

$$\Delta K(N) = \Delta K_0 + \beta_1 e^{-\beta_2 N} \tag{4}$$

Deriving and plotting Eq. (3) towards ΔK points of Eq. (4), a "regularised" da/dN-ΔK curve can be obtained. From this, the threshold SIF at a conventional crack growth rate of $1 \cdot 10^{-7}$ mm/cycle is derived.

The dependence of the threshold SIF on the load ratio can be analysed by applying some classical models. For example, the linear fit proposed by Barson (1974), Eq. (5), and the power fit proposed by Klesnil and Lukáš (1972), Eq. (6):

$$\Delta K_{th} = A - BR \tag{5}$$

$$\Delta K_{th} = D(1 - R)^{\gamma} \quad (6)$$

are applied in this study, with opportune fitting constants A, B, D and γ.

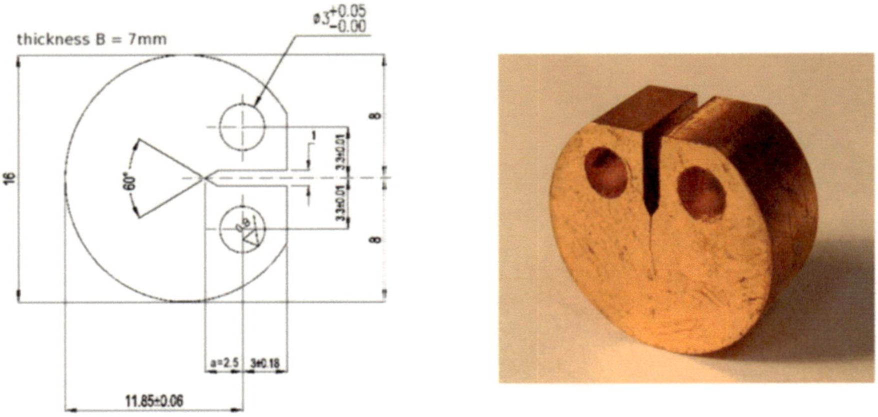

Fig. 9. The Disk Shaped CT specimen adopted for the experimentation; a propagated crack emanating from the notch is visible on the right-hand picture.

4. Fatigue crack growth resistance

4.1 Stable crack growth regime

Results of the FCG tests are now presented in the form of traditional log-log (da/dN; ΔK) diagrams. For the sake of clarity, only one propagation curve per R-ratio is plotted.

In Fig. 10, the crack propagation in the stable regime is plotted. Other data from the literature are plotted for comparison. In particular, these are: i) FCG curve at R = 0.25 of pure UFG Cu ECAPed by 4Bc passes with d_G = 270 nm (Cavaliere, 2009); ii) FCG curve at R = 0.5 of UFG copper produced by 4Bc passes and d_G = 300 nm (Vinogradov, 2007); iii) FCG curve at R = 0.5 of UFG copper produced by 16A ECAP passes and d_G = 300 nm (Vinogradov, 2007); iv) FCG curve at R = 0.5 of CG copper with d_G = 15 μm (Murphy, 1981).

From the analysis of the data some considerations can be drawn: 1) applied ΔK ranges between 6÷45 MPa√m, and related FCG rates between $6 \cdot 10^{-7} \div 2 \cdot 10^{-3}$ mm/cycle, the transition from stage I to stage II is continuous, and the conventional metals points in stage II can be fitted by the classical Paris propagation law:

$$da/dN = C(\Delta K)^m \quad (7)$$

2) the load ratio at stage II influences the propagation mechanism (Fig. 10 right); 3) in stage II the propagation curves intersect around the growth rate of 10^{-5} mm/cycle.

When comparing these with previous results, one can notice that: 1) a slower FCG characterises the present UFG copper for all R-ratios; 2) the slope of the curves in the Paris regime is comparable with data from the literature, i.e. a similar crack growth mechanism is found; 3) in stage II, the present UFG Cu shows higher FCG resistance than the CG counterpart.

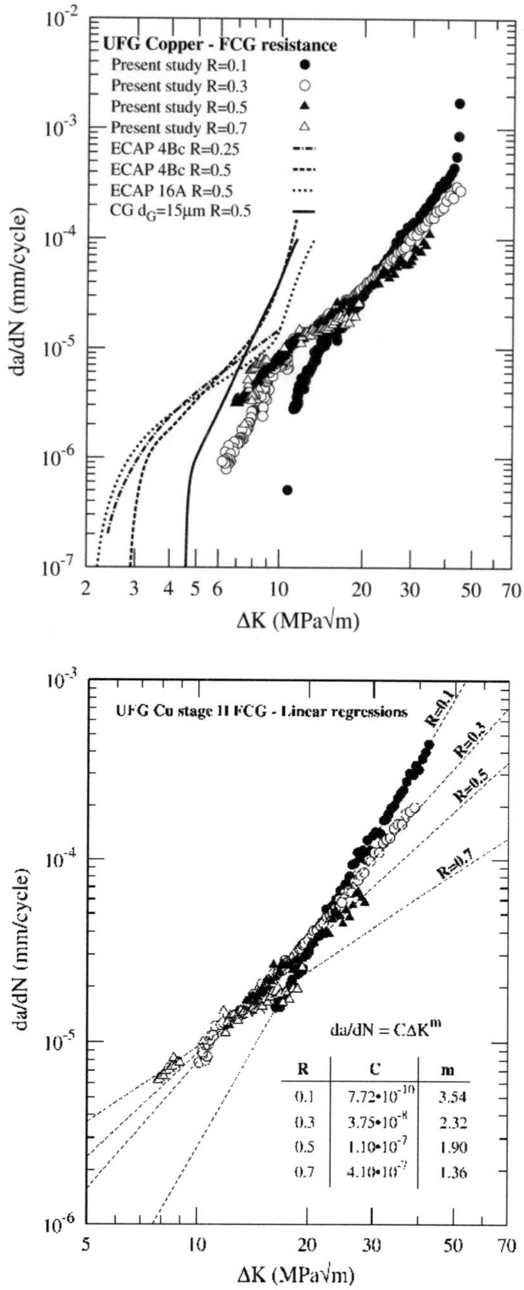

Fig. 10. FCG curves (left); linear regressions of stage II FCG rate (right).

4.2 Threshold regime (stage I)

In metals and alloys, the threshold FCG regime is usually defined by an FCG rate of around 10^{-7} mm/cycle. Since in the present tests only at R = 0.1 did the crack growth rate go under 10^{-6} mm/cycle, the threshold SIFs ΔK_{th} have been determined by the method reported in section 3.2. The results are summarised in Tab. 3, and compared with ΔK_{th} values taken from the literature.

The analysis of the FCG curves near the threshold regime and values of extrapolated ΔK_{th} highlights the following. The load ratio in UFG crack propagation influences the threshold FCG regime, since the higher the *R*, lower the ΔK_{th}, as one can evince from Fig. 10. Moreover, the threshold SIFs are higher than values found in the literature for the same class of UFG copper, as indicated in Tab. 3 and illustrated in the trend of the *C* coefficient in Eq. (7) (see Fig. 11). Finally, if compared with the FCG behaviour of annealed and cold worked conventional Cu alloys, the present UFG microstructure shows a higher threshold resistance to *R*-ratios 0.1 and 0.3, but lower resistance when the *R*-ratio increases, see Fig. 12; this result is partially in contradiction to other investigations on different purity levels.

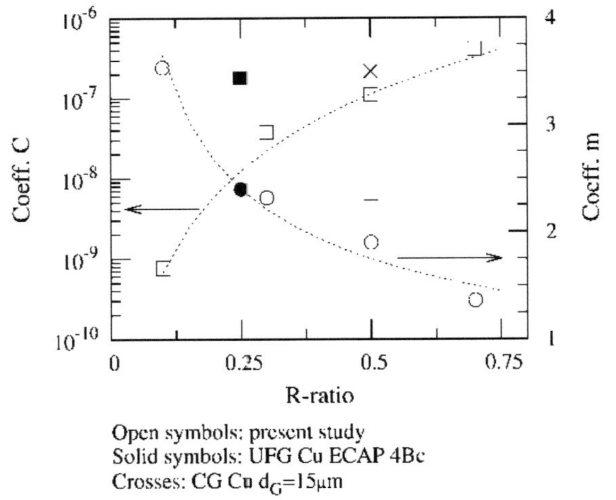

Fig. 11. Paris' law coefficient of Eq. (7) as a function of load ratio *R*.

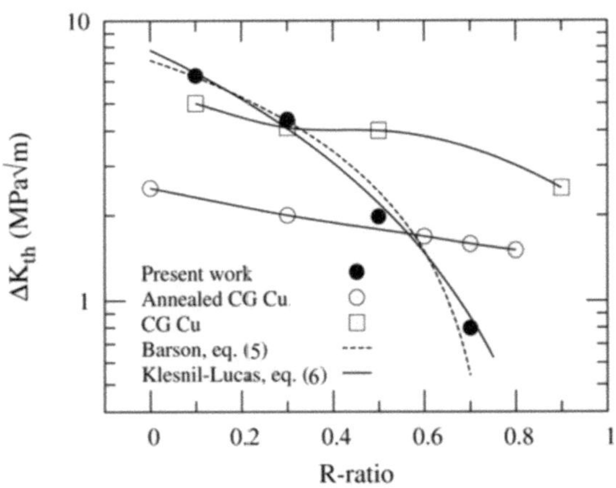

Fig. 12. Threshold stress intensity factors as a function of *R*-ratio.

Results of the threshold have been elaborated by Eq. (5) and Eq. (6) with the following fitting parameters: A = 9.45 MPa√m; B = 7.16 MPa√m; D = 7.79 MPa√m; γ = 1.82.

R-ratio	Present UFG Cu, ECAP 8Bc	UFG Cu, ECAP 4Bc	UFG Cu, ECAP 4Bc	UFG Cu, ECAP 16A
0.1	6.3	–	4.4	2.7
0.3	4.4	2.3 (R=0.25)	–	–
0.5	2.0	–	–	–
0.7	0.8	–	–	–

Table 3. Present ΔK_{th} values and data from literature.

4.3 Discussion of results

The experimental results on FCG resistance identify two important aspects: 1) the FCG resistance of the present UFG copper is higher than that of the previously tested ECAPed copper alloys, in stage II and also in the threshold propagation regime: this can be seen in the graphs of Fig. 10 and from Tab. 3; 2) load ratio influences the threshold stress intensity factor and the mechanism of stage II crack propagation. These two aspects will now be taken into consideration.

In order to explain the relatively high crack propagation resistance of the present UFG copper, its fatigue resistance must be analysed. It has been recently shown that the fatigue strength of UFG copper with low purity is higher than that of conventional copper by a factor of 2. In particular, the copper used in Kunz et al. (2006) had the same chemical composition and ECAP processing of the present material. Its higher fatigue resistance has been justified by demonstrating the stability of the bulk microstructure during cycling, due to the stable dislocation structure and to the presence of impurities and precipitates.

The grain structure within plastic zone around the cracks was shown to differ from outside the plastic zone: the grains were found to be markedly elongated, but their size was shown to be preserved. Also, in comparison with the CG structure, a small grain size can potentially result in more homogeneous deformation, which can retard crack nucleation by reducing stress concentrations and ultimately raise the fatigue limit of the UFG structure. This has been demonstrated by other studies on ECAPed copper structures on low and high cycle fatigue (Hanlon et al., 2005; Estrin & Vinogradov, 2010).

Moreover, it can be thought that the interaction between a propagating crack and the GBs structure can produce retardation in the growth rate. In fact, in most planar slip materials, GBs provide "topological obstacles to the slip" (Vasudevan et al., 1997). This phenomenon has already been noticed and theoretical models on the crack-boundaries interaction developed, with the support of experimental evidence (Holzapfel et al., 2007; Zhai et al., 2000). In these studies, it has been shown that due to the crack-precipitate interaction at the GBs, the crack develops steps on the crack plane while bypassing the precipitates. The result is a fatigue crack retardation and deflection at a GB that, by leading to an increase in the free crack surface, produces a significant suppression of the crack propagation rate. This topological factor can be critical in the FCG behaviour of UFG metals, if one considers the huge number of GBs generated by the grain refinement process.

The analysis of the effect of the *R*-ratio on the propagation behaviour in stage II is more complicated. As can be clearly seen from Fig. 10, at a constant FCG rate propagation, a higher R-ratio produces a slower growth rate, as if the material becomes more insensitive to the crack. This trend is not usual for polycrystalline metals, but very similar behaviour can be observed in the propagation curves of Cavaliere (2009), in which the crack growth resistance of UFG Cu alloys with increasing ECAP passes (i.e. grain refinement) are reported: going from two CG structures with 11 and 15 μm, towards UFG structures with passes 4Bc, 12Bc, 16A and 16Bc, where a lower threshold but higher stage II resistances are constantly found. The same trend is noticed in the present material when the R-ratio increases.

In order to rationalize the influence of load ratio, a crack closure approach has been attempted during the elaboration of the experimental results. The Adjusted Compliance Ratio (ACR) model (Donald, 1997) has been chosen. Depending on the ductility, FCG rate and environmental effects, numerous causes can be responsible for anticipated crack closure; among them, residual plasticity and roughness due to a tortuous crack path, characterise the so-called plasticity-induced crack closure (PICC) and roughness-induced crack closure (RICC), respectively. PICC and RICC are common mechanisms in ductile metals; PICC is mainly related to the residual plastic deformation in the steady-state FCG regime, while at threshold closure RICC is favoured by microstructural asperities of the fracture surfaces. The ACR method is based on the hypothesis that the effectively applied ΔK at the crack tip, namely ΔK_{eff}, is proportional to the strain magnitude, or to the crack tip opening displacement (CTOD), which is defined as:

$$CTOD = \frac{\left(1-\nu^2\right)K_{\max}^2}{2\sigma_y E} \tag{8}$$

The effective SIF is then calculated by correcting the applied SIF by a parameter (the ACR parameter) defined as follows:

$$\Delta K_{eff} = \Delta K \cdot ACR \tag{9a}$$

$$ACR = \frac{C_s - C_i}{C_0 - C_i} \tag{9b}$$

where C_s, C_0 and C_i are the specimen secant compliance, the compliance above the opening load and the compliance prior the initiation of the crack, respectively.

The elaboration of experimental points by this model is depicted in Fig. 13. Two main observations arise: 1) propagation curves almost overlap in stage II, and run parallel; 2) points below the $1.5 \cdot 10^{-5}$ mm/cycle show poor physical sense. It can be concluded that the ACR model rationalises the R-ratio effect in FCG stage II with discrete approximation, while it fails when applied to the threshold regime, where, as already known, the role of the microstructure achieves greater importance. This result is in partial accordance with FCG tests made on conventional CG copper, where it has been observed that by decreasing the grain size the threshold ΔK increases, suggesting that in copper crack tip plasticity considerations are more important in determining the threshold values than crack closure effects (Marchand et al., 1988).

Due of the peculiar micro-scale grain structure of UFG copper, the extension of the "process area" around the crack tip where plastic deformation concentrates can be investigated. According to Irwin, the size of monotonic and cyclic plastic zones at the crack tip can be estimated by Eq. (10) and Eq. (11) respectively:

$$r_p = \frac{1}{3\pi}\left(\frac{K_{max}}{\sigma_y}\right)^2 \tag{10}$$

$$r_{pc} = \frac{1}{3\pi}\left(\frac{\Delta K}{2\sigma_y}\right)^2 \tag{11}$$

where σ_y is the yield stress.

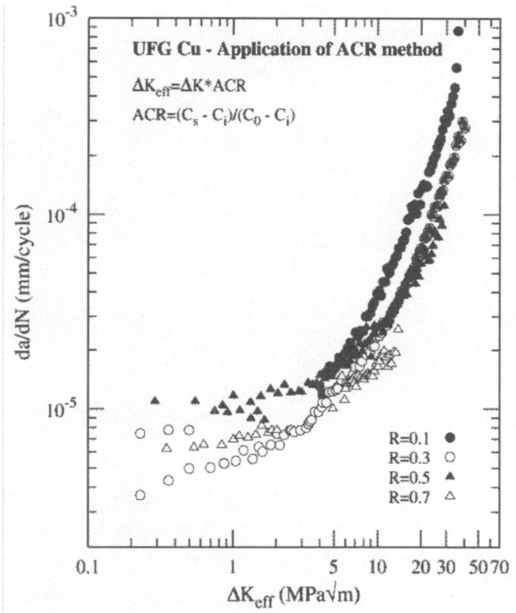

Fig. 13. Crack closure analysis by the application of the ACR model.

Results of this elaboration are shown in Fig. 14, where r_p and r_{pc} trends are reported as a function of the R-ratio, in respect to the high FCG rate ($2 \cdot 10^{-5}$ mm/cycle) and near-threshold FCG rate ($2 \cdot 10^{-7}$ mm/cycle). The same graph depicts the CTOD as calculated by Eq. (8). Fig. 14 illustrates that: 1) monotonic and cyclic plastic zones are always much wider than the characteristic microstructural dimensions (the average grain size d_G); 2) r_p and r_{pc} depend on the R-ratio in an opposite manner as a function of the FCG rate; 3) the CTOD is smaller when compared to r_p and r_{pc}. It other words, if the propagation of a fatigue crack is seen as the result of the accumulation of irreversible plastic deformation at the crack tip, in UFG copper this process involves one/two tenths of grains at a low FCG rate, but a large number of grains, up to some hundreds, at a high FCG rate.

In such a situation, one can expect a microstructure-dependent propagation mechanism at only very low FCG rates, as already observed for this material. This can partially explain the inadequacy of the ACR closure method when applied to the threshold regime. Fig. 14 also shows that at high FCG rates the cyclic plastic zone does not depend on the *R*-ratio, or, equivalently, on the applied average SIF calculated by Eq. (12):

$$\Delta K_{AVG} = \frac{1+R}{1-R}\Delta K \tag{12}$$

This indicates that for an increasing ΔK_{AVG}, the monotonic plastic zone expands, while the cyclic zone stabilises, being the microstructure able to resolve the external load.

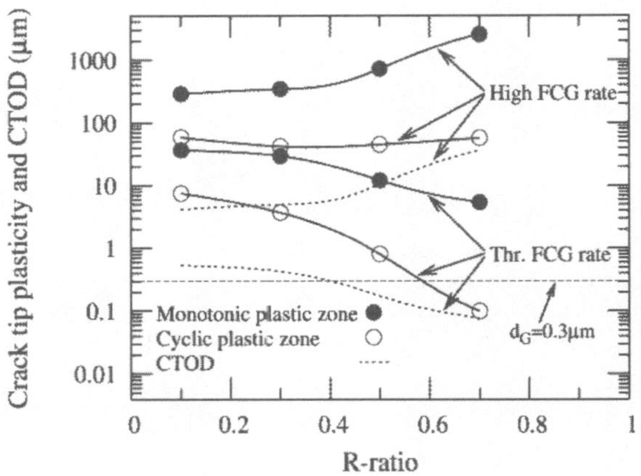

Fig. 14. Crack tip opening displacement (CTOD) and monotonic and cyclic plastic zones ahead of the crack tip calculated by Eqs. (8), (10) and (11) at $da/dN = 2 \cdot 10^{-7}$ mm/cycle and $da/dN = 2 \cdot 10^{-5}$ mm/cycle.

5. Fractographic analysis

The study of crack path and fracture surfaces is now presented to complete the analysis of the previously discussed FCG behaviour. The crack path morphology can be analysed in terms of fracture roughness R_v, measured on SEM images of cracked specimens. Roughness, that is a quantitative measure of the crack tortuosity, has been calculated by a simple graphical method, based on the superimposition of a grid on the digital image of a specimen profile.

The method is based on the counting of the profile-grid intersections, P_i, along a chosen length, L, and then expresses R_v as:

$$R_v = \frac{y}{L}\sum_i P_i \tag{13}$$

This simple technique gives a mono-dimensional estimation of roughness. R_v is calculated at different ranges of applied ΔK, for *R*-ratios equal to 0.3, 0.5 and 0.7. Results of the elaborations are presented in graphical form in Fig. 15. Crack tortuosity is higher when the driving force increases, but in practice it does not depend on the *R*-ratio. This conclusion confirms the insensitivity of the cyclic plastic zone with respect to the *R*-ratio at a given load range as already evidenced. Furthermore, it can be stated from Fig. 15 that R_v is only of the order of magnitude of the grain size d_G at the near-threshold regime. It means that the mechanism of fracture may only be microstructure-dependent during the threshold regime.

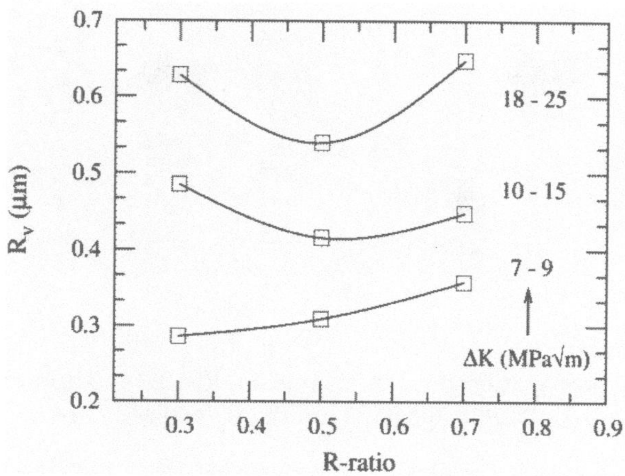

Fig. 15. Roughness of crack profile as a function of ΔK and R-ratio applied.

The fractographic analysis of the surfaces results a valuable tool to investigate the fracture mechanism and its dependence on load ratio. Some images of the crack path on the specimen profiles and of the fracture surface acquired by the SEM microscope are introduced in Figs. 16, 17 and 18.

Figs. 16 and 17 show the crack propagation profile at R = 0.3 and 0.7 respectively, with an increasing ΔK applied from left to right. Indications of the average SIF calculated by eq. (12) are also given. Looking at the profile morphology, an increasing tortuosity and fragmentation with an average ΔK can be seen for both R-ratios. Figs. 16(c) and 17(c), which correspond to the highest loads, show bifurcation, multi-cracking and branching of the main crack. Several small secondary cracks can be clearly seen in the perpendicular and parallel directions with respect to the main crack path direction, even at some distance from it. A very similar mechanism has been already been observed in commercially pure UFG Ti, in UFG Cu and, in AA6063 aluminium alloy.

A multi-cracking phenomenon indicates the poor capacity of the dislocation mechanism to generate around the crack tip and in the plastic wave of the hardened structure. Microcracks are generated to accommodate excessive strains in the crack vicinity, decreasing the strain hardening capability due to severe grain refinement. On the other hand, the branching mechanism may be directly responsible for the insensitivity toward crack propagation found at the stable FCG rate. During stage II, the FCG rate diminishes when R-ratio (or ΔK_{AVG}) increases, i.e. when branching becomes more evident. Indeed, it has been shown that crack deflection or multi-cracking can enhance K_{max} by a factor of about 20-30% (Vasudevan et al., 1997). This could definitely explain the FCG behaviour noticed, otherwise it would be very difficult to rationalise it with a crack closure approach.

Fig. 18 shows the fracture profile and the fracture surface near the threshold regime, at R = 0.3 and R = 0.5, respectively. The fracture morphology of Fig. 18(a), taken when the cyclic plastic zone at the crack tip was about 5-8 times the grain size, provides evidence of an intergranular mechanism, justifying the high value of ΔK_{th} found. The SEM image of Fig. 18(b) shows relatively long, straight-line (secondary) microcracks, perpendicular to the direction of crack growth. This is coherent with the previous observations, and indicates a rather brittle, intergranular micro-mechanism of propagation.

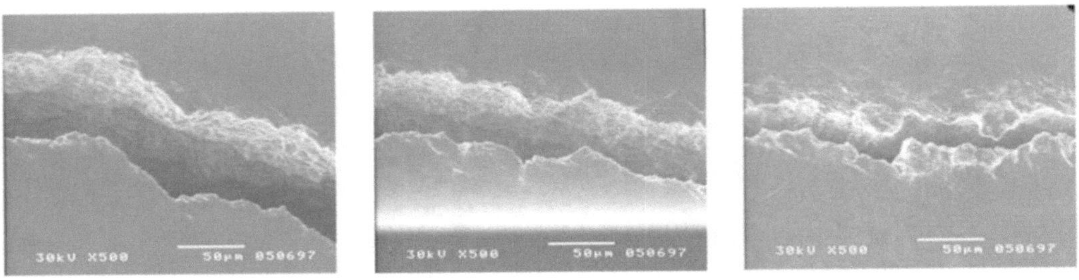

(a) (b) (c)

Fig. 16. Stage II crack propagation profiles at constant R = 0.3 and applied[average] ΔK equal to: (a) 8.6[16.0] MPa√m; (b) 11.2[20.8] MPa√m; (c) 41.0[76.1] MPa√m.

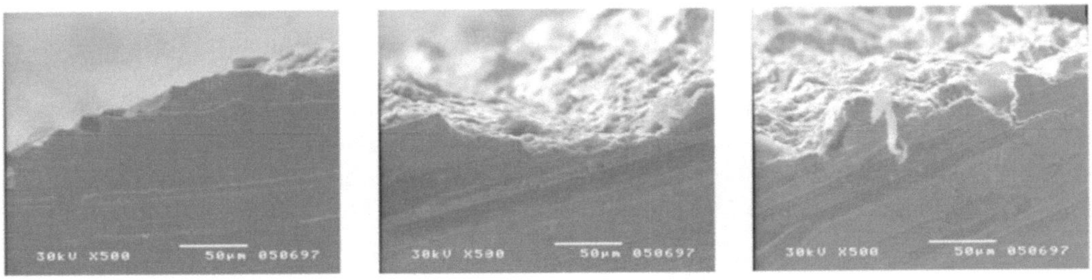

(a) (b) (c)

Fig. 17. Stage II crack propagation profiles at constant R = 0.7 and applied[average] ΔK equal to: (a) 8.1[45.9] MPa√m; (b) 12.3[69.7] MPa√m; (c) 18.7[106.0] MPa√m.

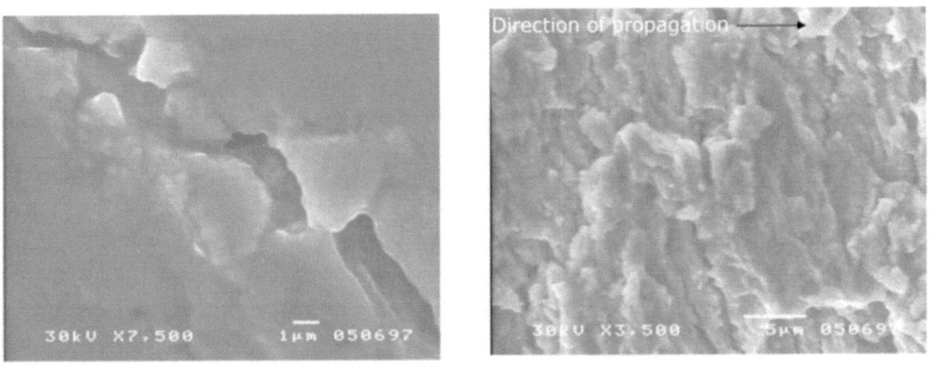

(a) (b)

Fig. 18. Crack paths during the threshold regime: (a) intergranular fracture at R = 0.3; (b) fracture surface morphology at R = 0.5.

6. A model of fatigue crack growth

The ACR model previously discussed is demonstrated as being adequate for interpreting the closure of UFG copper in a stable crack growth (microstructure-independent) regime.

The analysis of the SEM images of a typical crack path such as that reported in Fig. 19, induced the analyst to attempt a model of crack-deflection-induced closure. In fact, the image shows how the crack periodically deflects following a "zig-zag" path, which causes a premature closure mechanism. It's worth to notice that the direction of main propagation corresponds to the shear plane of the last ECAP passage.

The model proposed by Suresh (1985) has been applied. With the simple geometrical considerations shown in Fig. 19, this model calculates the reduction in the driving force for propagation throughout an effective ΔK_{eff} and a corresponding effective crack propagation rate, as specified in Eq. (14) (see the scheme of Fig. 20 for the symbols):

$$\frac{\Delta K_I}{\Delta K_{eff}} = \left(\frac{D\cos^2(\theta/2)+S}{D+S}\right)^{-1}\left(1-\sqrt{\frac{\chi\tan\theta}{\chi\tan\theta+1}}\right)^{-1}$$
$$\left(\frac{da}{dN}\right)_{eff} = \left(\frac{D\cos\theta+S}{D+S}\right)\left(\frac{da}{dN}\right)_L \tag{14}$$

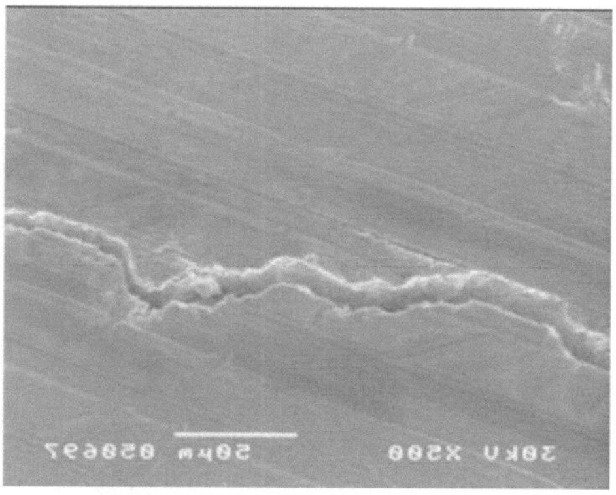

Fig. 19. Propagation path with crack deflections at R = 0.3 and applied ΔK = 20MPa√m.

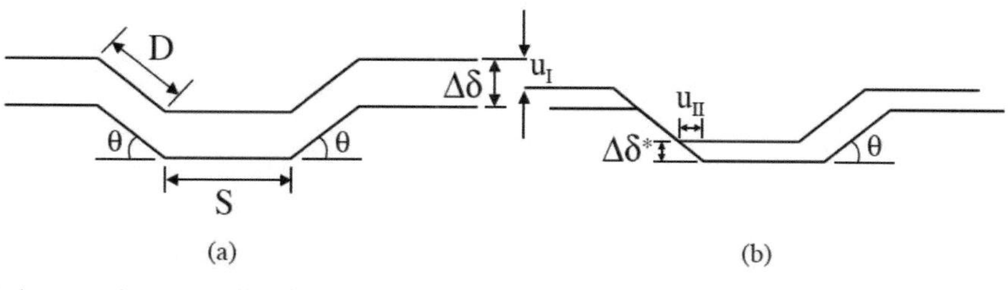

Fig. 20. Scheme of a periodically deflected crack: (a) opened state at the peak load: S is the straight length, D the deflected length, θ the deflection angle, $\Delta\delta$ the CTOD; (b) at the first point of contact upon unloading: u_I and u_{II} are the surface mismatches, $\Delta\delta^* = u_{II}\tan\theta$ is the closure CTOD.

The elaboration of the experimental points via this model is depicted in Fig. 21. Propagation points at the near-threshold regime are well interpreted by the model of Eq. (14), fitted with the following parameters: $D = S$ = 30 μm; θ = 26°; χ = 0.09 at R = 0.1, χ = 0.0125 at R = 0.3, χ = 0.0083 at R = 0.5, χ =0.0 at R = 0.7. It is clear that the ACR model rationalises the R-ratio effect in FCG stage II with a discrete approximation, while it fails when applied to the threshold regime, where evidently the role of the microstructure is of greater importance.

On the other hand, the model of the deflection of the crack rationalises well the closure mechanism occurring when the growth rate is very low. The mismatch parameter χ, which measures the level of crack closure, diminishes when the load ratio (or K_{max}) increases; this is due to the increase in fracture surface separation with an increasing R. The adopted values of the mismatch parameter u_I and u_{II}, however, indicate that a relatively few number of grains are involved in the process at a slow crack growth rate.

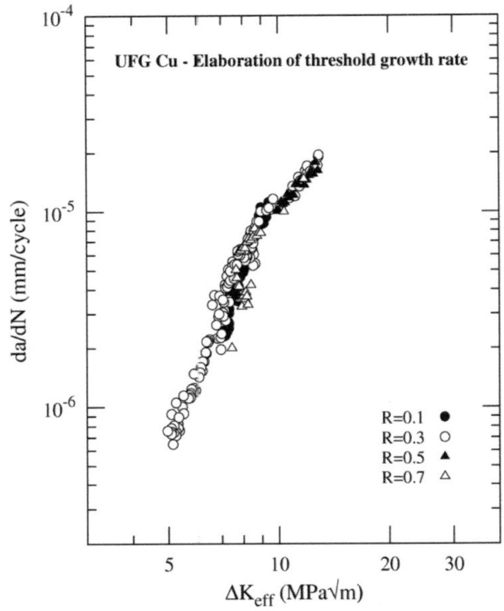

Fig. 21. Crack closure elaboration at stage II (left) and stage I (right).

The mechanism of propagation at high crack growth rates, and the interrelationship with the ultrafine structure remains to be fully investigated. As already evidenced, the fractographic analysis conducted on several crack surfaces from the SEM microscope, revealed an increasing tortuosity and fragmentation of the crack with an average ΔK, over all the R-ratios. The profile morphology at the highest loads shows bifurcation, multi-cracking and branching of the main crack, and several small secondary cracks appearing in perpendicular and parallel directions with respect to the main crack direction, and even at some distance from it. Microcracks generating to accommodate excessive strains in the crack vicinity cause a decrease in the strain hardening capability due to severe grain refinement. Nevertheless, the branching mechanism may be directly responsible for the insensitivity toward crack propagation found at the stable FCG rate. Indeed, during stage II propagation, the FCG rate diminishes when the R-ratio (i.e. K_{max}) increases, in other words when branching becomes more evident. Effectively, crack branching would reduce the mode I crack driving force and has been known to play a significant role in crack retardation. Recently, in a work on finite element modelling of fatigue crack branching, it has been shown that the near-tip SIF range was significantly reduced due to crack branching

(Meggiolaro et al., 2005). The crack tip stress shielding introduced by crack branching could be considered to be the reason for the lower crack growth rates observed, definitely explaining the FCG behaviour of the UFG structure witnessed.

7. Conclusions

The experiments have shown that the resistance to the crack propagation in ultrafine-grained copper alloy is not only influenced by the peculiar microstructure and the technological process employed to obtain it, but also by hardening conditions and boundary impurities.

Different crack growth mechanisms are individuated, depending on the crack growth rate (threshold or Paris' regime). In particular, elaboration of the experimental data shows that: 1) a large number of grains are involved in the propagation process; 2) a plastic induced crack closure mechanism is the most probable closure mechanism at high driving force, while a roughness induced crack closure mechanism dominates near the threshold growth rate: this is consistent with the average material grain size.

Finally, the increasing insensitivity toward the fatigue crack propagation shown during the Paris' regime when the *R*-ratio increases, indicates a mechanism of apparent toughening that can be explained, at least qualitatively, by considering the observed branching of the crack path along the ECAP shear planes.

However, the mechanism of the crack propagation in UFG Cu, and its relation to the microstructure and its changes, up until now have not been sufficiently studied. Further goal-directed investigations are necessary to obtain further knowledge on this issue.

8. Acknowledgements

The author wishes to thank Prof. Ludvik Kunz from the Institute of Physics of Material in Brno, Czech Republic, for providing the material for this research activity, and for the fruitful discussions.

9. Nomenclature

a	crack length (mm)
W	specimen width (mm)
B	specimen thickness (mm)
P	tensile applied load (N or kN)
$C_0 \div C_4$	constants relative to the DSCT geometry
N	number of load cycles
da/dN	crack growth rate (mm/cycle)
K	Stress Intensity Factor (SIF) (MPa√m)
K_a	applied SIF
ΔK	Stress Intensity Factor amplitude = $K_{max}-K_{min}$ (MPa√m)
ΔK_{th}	Threshold Stress Intensity Factor (MPa√m)

K_{op}	Stress Intensity Factor necessary to open a crack (MPa√m)
R	load ratio = K_{min}/K_{max}
C, m	Paris propagation law constants
d_G	average grain size (μm)
$\kappa_{1,2}$	interpolating constants
$\beta_{1,2}$	interpolating constants
A, B	fitting parameters
D, γ	fitting parameters
C_s, C_0, C_i	specimen compliances: secant; above K_{op}; prior the initiation of crack
ΔK_{eff}	Effective Stress Intensity Factor = K_{max}–K_{op} (MPa√m)
$(da/dN)_L$	growth rate of a linear undeflected crack (mm/cycle)
S, D, θ, $\Delta\delta$, $\Delta\delta^*$	geometrical parameter of crack deflection
u_I, u_{II}	sliding displacement at the crack closure
χ	mismatch parameter = u_{II}/u_I

10. References

Barson J. Fatigue behaviour of pressure-vessel steels. Welding Res Council Bull 1974;194.

Cavaliere P. Fatigue properties and crack behavior of ultra-fine and nanocrystalline pure metals. Int J Fatigue. 2009;31(10):1476-1489.

Collini L. Fatigue crack growth in ECAPed commercially pure UFG copper. Proc Eng. 2010a;2(1):2065-2074.

Collini L. Fatigue crack growth resistance of ECAPed ultrafine-grained copper. Eng Fract Mech. 2010b;77:1001-1011.

Donald J.K. Introducing the compliance ratio concept for determining effective stress intensity. Int J Fatigue. 1997;19(1), S191-S195.

Estrin Y., Vinogradov A. Fatigue behaviour of light alloys with ultrafine grain structure produced by severe plastic deformation: an overview. Int J Fatigue. 2010;32(6):898-907.

Goto M, Han S.Z., Kim S.S., Ando Y., Kawagoishi N. Growth mechanism of a small surface crack of ultrafine-grained copper in a high-cycle fatigue regime. Scripta Mater. 2009;60:729-732.

Hanlon T., Tabachnikova E.D., Suresh S. Fatigue behavior of nanocrystalline metals and alloys. Int J Fatigue 2005;27:1147-1158.

Holzapfel C., Schäf W., Marx M., Vehoff H., Mücklich F. Interaction of cracks with precipitates and grain boundaries: understanding crack growth mechanisms through focused ion beam tomography. Scripta Mater. 2007;56:697-700.

Höppel H.W., Kautz M., Xu C., Murashkin M., Langdon T.G., Valiev R.Z., Mughrabi H. An overview: Fatigue behaviour of ultrafine-grained metals and alloys. Int J Fatigue. 2006;28:1001-1010.

Horky J., Khatibi G., Weiss B., Zehetbauer M.J. Role of structural parameters of ultra-fine grained Cu for its fatigue and crack growth behaviour. J Alloy Compd. 2011;509S:S323-S327.

Klesnil M., Lukáš P. Effect of stress cycle asymmetry on fatigue crack growth. Mater Sci Eng 1972;9:231-40.

Kozlov E.V., Zhdanov A.N., Popova N.A., Pekarskaya E.E., Koneva N.A. Subgrain structure and internal stress fields in UFG materials: problem of Hall–Petch relation. Mater Sci Eng A. 2004;387–389:789-794.

Kunz L., Lukáš P., Svoboda M. Fatigue strength, microstructural stability and strain localization in ultrafine-grained copper. Mater Sci Eng A. 2006;424:97-104.

Lugo N., Llorca N., Cabrera J.M., Horita Z. Microstructures and mechanical properties of pure copper deformed severely by equal-channel angular pressing and high pressure torsion. Mater Sci Eng A. 2008;477(1-2):366-371.

Lukáš P., Kunz L. Mechanisms of near-threshold fatigue crack propagation and high cycle fatigue in copper. Acta Technica CSAV. 1986;4:460-488.

Lukáš P., Kunz L., Knésl Z. Fatigue crack propagation rate and the crack tip plastic strain amplitude in polycrystalline copper. Mat. Sci. Eng. 1985;70:91-100.

Lukáš P., Kunz L. Svoboda M. Fatigue mechanism in ultrafine-grained copper. Kovove Mater. 2009;47:1-9.

Marchand N.J., Baïlon J.-P., Dickson J.I. Near-threshold fatigue crack growth in copper and alpha-brass: grain-size and environmental effects. Metall Mater Trans A. 1988;19(10):2575:2587.

Meggiolaro M.A., Miranda A.C.O., Castro J.T.P., Martha L.F. Crack retardation equations for the propagation of branched fatigue cracks. Int J Fatigue. 2005;27:1398-1407.

Meyers M.A., Mishra A., Benson D.J. Mechanical properties of nanocrystalline materials. Prog Mater Sci. 2006;51:427-556.

Mughrabi H., Höppel H.W., Kautz M. Fatigue and microstructure of ultrafine-grained metals produced by severe plastic deformation. Scripta Mater. 2004;51:807-814.

Murphy M.C. The engineering fatigue properties of wrought copper. Fatigue Eng Mater Struct. 1981;4(3):199-234.

Suresh S. Fatigue crack deflection and fracture surface contact: micro-mechanical models. Metall Trans. 1985;16A:249-260.

Thompson A.W., Backofen W.A. The effect of grain size on fatigue. Acta Metal. 1971;19(7):597-606.

Valiev R.Z. Structure and mechanical properties of ultrafine-grained metals. Mater Sci Eng A. 1997;234-236:59-66.

Valiev R.Z., Islamgaliev R.K., Alexandrov I.V. Bulk nanostructured materials from severe plastic deformation. Progr Mat Sci. 2000;45:103-189.

Valiev R.Z., Langdon T.G. Principles of equal-channel angular pressing as a processing tool for grain refinement. Prog Mater Sci. 2006;51:881-981.

Vasudevan A.K., Sadananda K., Rajan K. Role of microstructures on the growth of long fatigue cracks. Int J Fatigue. 1997;19(1):S151-S159.

Vinogradov A. Fatigue limit and crack growth in ultra-fine grain metals produced by severe plastic deformation. J Mater Sci. 2007;42:1797-1808.

Wei W., Chen G. Microstructure and tensile properties of ultrafine grained copper processed by equal-channel angular pressing. Rare Metals. 2006;25(6): 697-704.

Xu C., Wang Q., Zheng M., Li J., Huang M., Jia Q., Zhu J., Kunz L., Buksa M. Fatigue behavior and damage characteristic of ultra-fine grain low-purity copper processed by equal-channel angular pressing (ECAP). Mater Sci Eng A. 2008;475(1-2):249-256.

Zhai T., Wilkinson A.J., Martin J.W. A crystallographic mechanism for fatigue crack propagation through grain boundaries. Acta Mater. 2000;48:4917-4927.

Part 3

Archaeometallurgy of Copper Alloys

7

Bronze in Archaeology: A Review of the Archaeometallurgy of Bronze in Ancient Iran

Omid Oudbashi[1], S. Mohammadamin Emami[1,2] and Parviz Davami[3]
[1]*Faculty of Conservation, Art University of Isfahan,*
[2]*Institut fur Bau- und Werkstoffchemie, Universitaet Siegen,*
[3]*Faculty of Materials Science and Engineering, Sharif University of Technology,*
[2]*Germany*
[1,3]*Iran*

1. Introduction

The history of metals and metallurgy is rooted in the history of civilizations as the "Archaeometallurgy" and has been a subject of great interest for over a century. Due to the relatively good preservation of metallic goods and the modern values related to metals, metal artefact typologies often served as the very basis for prehistoric sequences during the late 19th and early 20th centuries. In many ways, it was V. Gordon Childe who placed metallurgical technology at the front, arguing as he did for the roles of "itinerant metal smiths" and bronze production in the rise of social elites and complex societies. Childe was also one of the first to systematically argue for the transmission of metallurgy from the Near East to the Eurasia (Thornton & Roberts, 2009). On the other hand, many of the artefacts which excavated, as well as some of the metallurgical talent being practiced are standing examples that depict the superior metallurgical skills used by human. Archaeometallurgical investigations can provide evidence about both the nature and level of mining, smelting and metalworking trades, and support understanding about structural and technical evidences. Such evidence can be essential in understanding the economy of a settlement, the nature of the industry and craft, the technological capabilities of its craftsmen as well as their cultural relations. In order to achieve such data, it is obvious that archaeometallurgical discipline has considered at each stage of archaeological and historical investigations in the field of ancient metal working.

The development of metallurgy on the Iranian Plateau has been a topic of interest to both archaeologists and scientists for many years because of the remarkable history of the metallurgical activities in Iran (such as usage of native copper in the 7th millennium BCE and smelting of copper ores by the late 6th millennium BCE) and concerned the wide variety of the technologies, compositions, etc. Indeed, the rich and old history of the Iranian Plateau and the huge metallurgical and metal working remnants spread in various forms and different parts of this region have been an important source for archaeological and archaeometallurgical studies for many years, especially during the last decade (e.g., Arab & Rehren, 2004; Pleiner, 2004; Thornton & Rehren, 2007).

A significant number of prehistoric sites have been excavated and uncovered, especially during the second half of the 20th century in Iran. Many of them yield valuable information concerning ancient metallurgy and metal working in the Iranian plateau during prehistoric period. Various metal artefacts, moulds, slags, crucibles and other tools and materials have been recovered belonging to prehistoric metalworkers from a series of excavations in different geographical regions of Iran (Figure 1). These archaeological finds and their location in Iran have created information and materials to be studied in relation to different aspects of ancient Iranian metallurgy.

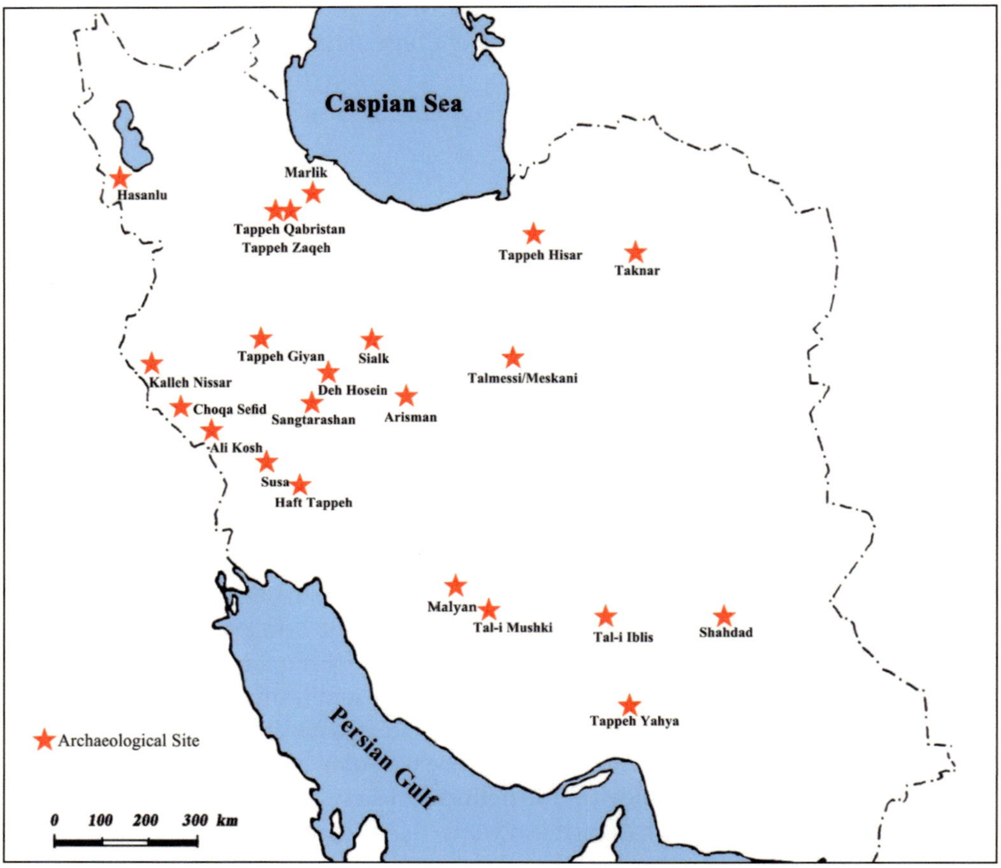

Fig. 1. Approximate location of important metallurgical sites mentioned in the text.

Ancient artefacts made by copper and its alloys belonging to prehistoric period, from the 7th millennium BCE (Thornton, 2009a) down to the 1st millennium BCE (Overlaet, 2006), discovered from different sites, are evidences for the ancient metallurgy of copper and copper alloys in Iran.

In the field of archaeometallurgical studies, there are many reports and papers about study of ancient ore mining, slags and other metallurgical remains, artefacts and so on between various archaeological investigations about Iran; such as works on Tappeh Yahya (Tepe Yahya) in southeast Iran (Heskel & Lamberg-Karlovsky, 1980; Heskel & Lamberg-Karlovsky 1986; Thornton et al., 2002; Thornton & Ehler, 2003; Thornton & Lamberg-Karlovsky, 2004);

Tappeh Hisar (Tepe Hissar), Northern Iran (Thornton 2009b); Godin Tappeh (Godin Tepe) (Frame, 2007) and Tel-i Iblis (Frame, 2004); Luristan Bronzes (e.g., Oudbashi et al., in Press; Fleming et al., 2005; Fleming et al., 2006; Moorey, 1964, 1969; Birmingham et al., 1964); Haft Tappeh (Haft Tepe), southwest Iran (Oudbashi et al., 2009); and ancient slags from Meymand, Kerman, southeast Iran (Emami & Oudbashi, 2008).

The aim of this paper is to review copper archaeometallurgy in Iranian Plateau, with a special regard to bronze (Cu-Sn) technology in prehistoric period (until mid-1st millennium BCE). Also, results of a comparative study on copper and bronze metallurgy in two ancient sites from western and southwestern parts of Iran are presented.

2. Chronology of Iran prehistory

Iran is one of the oldest areas with evidence for human life and settlement. There are many witnesses from human activities in Iranian Plateau from many ages. Human settlement starts in Iran from Paleolithic age and may belong to middle Paleolithic tradition (Table 1). We cannot be certain when the Middle Paleolithic began in Iran, but it has ended before 30,000 years ago. Also there are some remains of human living from upper Paleolithic period (terminated about 17,000 years age) and terminal Paleolithic (terminated about 10,000 years ago) in different areas of Iranian Plateau (Hole, 2008). After that, the Neolithic age has been happened about 10,000 years ago by starting and combination of agriculture (wheat, barley, lentils) and the raising of livestock (goats, sheep, cattle, pigs), formed the basis of the agricultural economy that has lasted until today and spread throughout the world.

This ended about 6500-7000 years ago. The earliest Neolithic occurred before the use of hand-made, chaff-tempered pottery which appeared around 8,500 B.P. The Neolithic ends with the appearance of new styles of pottery, generally with designs painted in black on a buff background (Hole, 2004). The next steps are the ages of metals, which are: the Chalcolithic (copper), Bronze, and Iron Ages. In Near Eastern archaeology, the Chalcolithic now generally refers to the "evolutionary" interval between two "revolutionary" eras of cultural development: the Neolithic, during which techniques of food production and permanent village settlement were established, and the Bronze Age, during which the first cities and state organizations arose (Henrickson, 1992a, 1985). In Iran the Chalcolithic dates between about 5500-3300 BCE and smelting of copper ores and casting metallic copper artefacts has been started in this period. During and after the Chalcolithic age, the usage of metal and various aspects of metallurgical technology developed (Tala'i, 2008b). As a result for developing human culture and technology, a period observed in which the rise of trading towns in Iran has been occurred, from ca. 3300 BCE to the beginning of the Iron Age, ca. 1500 BCE, that has been named as Bronze Age. As a metallurgical phenomenon, first observations about application of bronze (Cu-Sn) alloy have been occurred in early Bronze Age (Dyson Jr. & Voigt, 1989). The last step of Iran prehistory is the Iron Age. In Iran the term Iron Age is employed to identify a cultural change that occurred centuries earlier than the time accorded its use elsewhere in the Near East, and not to acknowledge the introduction of a new metal (Iron) technology. The Iron Age has been dated about 1500-550 BCE and classified in three levels: Iron Age I, II and III. Iron artefacts, in fact, were not be widespread in Iran until the 9th century BCE (the cultural period labeled Iron Age II), centuries after the phase designated as Iron Age I came into existence (Muscarella, 2006; Overlaet 2005). After Iron Age III, the Iranian Plateau entered to Ancient Persian dynasties, started with Achaemenids (550-330 BCE) and ended with Sasanians (240-641 AD).

Period	Categories	Date*	Characteristics	Metallurgy
Paleolithic	Lower	N.D.**	Early Stone Tools, Hunting, Cave Dwelling, Fire Discovery, No Evidence of "Art"	No evidences
	Middle	N.D.-30000		
	Upper	30000-17000		
	Terminal	17000-10000		
Neolithic	Aceramic	8000-7300	Introduction of Agriculture, Livestock and Villages, Fine Stone Tools, Early Potteries	Early usage of native copper
	Ceramic	7300[1]-5500		
Chalcolithic	Early	5500-5000	Progress in Agriculture, Permanent Villages, Widespread Painted Potteries, Advanced Pottery Kilns	Copper melting and copper ore smelting, Arsenical copper, Silver and Gold appearance
	Middle	5000-3800[2]		
	Late	3800-3300		
Bronze Age	Early	3000-2200	Painted Pottery, Early Towns, Long-Distance Trade	Appearance of early low tin-bronzes, Large scale metallurgical workshops, early brasses
	Middle	2200-1800		
	Late	1800-1500		
Iron Age	I	1500-1000	Local Traditions, Special Cemeteries, Grey Pottery,	Common use of bronze, appearance of Iron, common use of gold and silver,
	II	1000-800		
	III	800-550		
Elam	Proto-Elamite	3200-2800	Appearance of Writing (Proto-Elamite), First Dynasty in Iran, Under Influence of Mesopotamia, Significance Architecture, Seals, Cuneiform Inscriptions	Large scale use of copper and bronze, Iron appearance and usage, gold and silver in metal decoration
	Old Elamite	2200-1500		
	Middle Elamite	1500-1100		
	Neo Elamite	1000-640		

* The Paleolithic dates are before present and from Neolithic afterward, the dates are presented as BCE.

**N.D.: Not Determined.

[1] Ceramic Neolithic doesn't start in Northern and Eastern Iran until much later (ca. 6500-6000 BCE).

[2] This date is related to Zagros Highlands in Iranian Plateau, Please see: Henrickson, 1992b.

Table 1. A diagram showing chronology, characteristics and metallurgical events of Iranian Prehistory eras and Elamite period.

On the other hand, the plain of Khuzestan in the southwest Iran played a major role with the origin of urban societies in the Middle East besides Mesopotamia, which called Elam. The earliest evidence mentioning the country of Elam is from Mesopotamia and belongs to the 3rd millennium BCE. The Elamite region was not restricted to the plain of what is Khuzestan today but included wide parts of the Zagros Mountain to the North and East, as well as the region of Fars. Chogha Zanbil (Dur Untash), Haft Tappeh and Susa are the main cities that have been discovered from Elamite civilization. The Elamite period has divided to four specific categories: Proto, Old, Middle and Neo Elamite. This period is between last 4th/early 3rd millennium BCE and mid of 1st millennium BCE (Mofidi Nasrabadi, 2004; Potts, 1999).

3. Copper metallurgy in prehistoric Iran

3.1 Native copper

Copper is one of the most useful metals, and probably is the first metal that has been used for manufacturing different tools and artefacts. The characteristics of copper caused to use it for making jewelry in middle of Neolithic period. First application of copper to use as primary tools comes back to about 10 thousand years ago in the Near East and namely Iranian Plateau. It has already been mentioned that the first metal objects appear in southwest Iran, Deh Luran plain at the base of the Zagros Mountains which lies on a traditional route between Mesopotamia and the Susiana plain (Pigott, 2004a).

In fact, the first usage of copper for making an artefact in Iran may comes from the Neolithic site in southwestern Iran, namely Ali Kosh, where one piece of rolled bead of native copper has been found (Figure 2) (Smith, 1967; Moorey, 1969; Pigott, 2004a; Thornton, 2009a). This is recently dated to the late 8th/early 7th millennium BCE (Hole, 2000; Thornton, 2009a). Also, some witnesses about early copper finds have observed in Tappeh Zaqeh (North central Iran) and Choqa Sefid (Western Iran) (Pernicka, 2004; Bernbeck, 2004). Analyses of these metal artefacts have revealed that all were made of native copper (like Ali Kosh) with the application of heat treatment (annealing) on metal between various deformation steps in order to avoid work-hardening (Pernicka, 2004). Of course, implementation of metal usage occurred much later in the early-mid 6th millennium BCE, when native copper artefacts were utilized consistently in various parts of the Iranian Plateau such as Tal-i Mushki archaeological site in Marvdasht, southern Iran (Moorey, 1982; Thornton, 2009a).

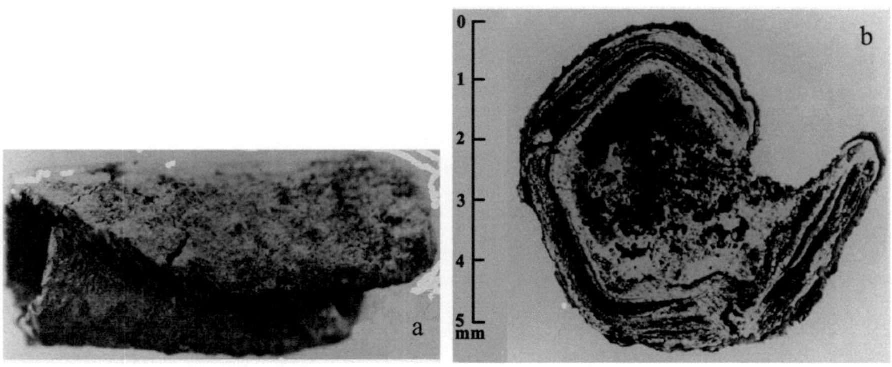

Fig. 2. a) Native copper rolled bead from Ali Kosh Neolithic site, western Iran (mid 7th millennium), b) Polished cross-section of the copper bead. Metal is corroded but the resulting corrosion products have preserved its original shape (Pigott, 2004a; Smith, 1967).

3.2 Copper casting and smelting

It has been assumed that the melting and casting native copper may have been the early stage of pyrometallurgy before ores smelting (Wertime, 1973). The first clear evidence of copper casting is distinguished in late 5th/early 4th millennium BCE (e.g., Level III at Sialk, central Iran) (Moorey, 1969). The 5th millennium BCE on the Iranian Plateau witnesses the transition from the use of pure native copper to the smelting of copper ores chosen for their natural impurities such as arsenic (Thornton et al., 2002). At the same time metal is more commonly used for tools previously made by bone and stone (Moorey, 1969). Arsenic has found even in first smelted copper artefacts and even earlier in native copper objects as accidental alloying element or impurity (Thornton 2009a). Although, the arsenical copper production may assume more as an important stage of casting process than producing an alloy (Tala'i, 2008b; Tala'i, 1996).

The Cu, As-bearing minerals contained in the native copper are very important because the early metalworkers in Iran began to melt the native copper in order to cast it, and it caused producing arsenical copper (accidental alloying). This phenomenon may also have occurred by melting native copper containing arsenides in crucible (Pigott, 2004a). Anarak economical resource area in central Iran is notable for copper casting because there are two large outcrops of native as well as arsenical copper. The orogeny zone of Talmessi and Meskani, is important according to high enrichments of copper arsenides, as algodonite ($Cu_{6-7}As$), enargite (Cu_3AsS_4) and domeykite (Cu_3As) (Pigott 2004a; Thornton et al., 2002). Another copper ore deposition with arsenic enrichments in literature is Taknar, closed to the metallurgical site, Tappeh Hisar, northern Iran. As a matter of fact, the use of furnace as evidences for copper smelting has been carried out in Tappeh Hisar (Pigott et al., 1982).

Metallurgy of copper has followed by the manufacturing of artefacts with arsenical copper from 4th millennium BCE. The copper extraction from different ore resources (furnace-based metallurgy) may occurs as next step of copper metallurgy. The copper ores smelting in furnace has begun in 5th millennium BCE by smelting oxidic ores such as cuprite or malachite, for example in Tal-i Iblis (southeast Iran), Tappeh Qabristan and Sialk (north-Central Iran), Tappeh Hisar (northeast Iran) and Susa (Southwest Iran) (Figure 3) (Thornton, 2009a; Thornton et al., 2002; Pernicka, 2004; Dougherty & Caldwell, 1966). One of the important sites in copper smelting is Arisman, Central Iran. Archaeological excavations executed in this site (with slag concentrations) cleared that extensive copper smelting took place at the site during the late 5th to the early 3rd millennium BCE (Pernicka, 2004; Chegini, et al., 2004).

Analytical results from various archaeological areas in the Near East and the eastern Mediterranean region have made clearly that an intentional arsenic-copper alloy was often, if not invariably, an important stage in the transition from cast copper to the use of a tin-copper alloy. First trying to make an alloy may be producing copper-arsenic alloy or arsenical copper in prehistory (Tala'i, 2008a; Scott, 2002; Thornton, 2010). This alloy became widespread in the Near East sometime in the second half of the 4th millennium BCE and might well have arisen from an accidental use of an arsenic-enriched copper ore (Moorey, 1969).

Fig. 3. A double axe made of arsenical copper from chalcolithic period (Susa I/II, middle of the 4th millennium BCE), Louvre Museum (Benoit, 2004).

As amount of lower than 2 percent may show that it has entered into composition as an impurity from copper ores such as tennatite, $(Cu,Fe)_{12}As_4S_{13}$ (Grey Copper) and this can be considered as accidental alloy (Coghlan, 1975). With high amount of As, Coghlan (1975) suggested that the intentional alloying may be occurred by three metallurgical procedures:

1. Contemporaneous smelting of copper oxide ores with realgar (As_4S_4) or orpiment (As_2S_3).
2. Use of ores with high As-content such as arsenopyrite or tennatite mixed with copper sulfides.
3. Adding realgar or orpiment to the melt.

On the other hand, Thornton et al (2009) suggest the speiss (iron-arsenide alloy) as a quasi-metallic material usually formed as an accidental by-product of copper or lead smelting. It has been produced to provide arsenic as an alloying component for arsenical copper.

3.3 Tin bronze

In archaeometallurgical literature, bronze is a copper alloy which mostly consists of tin. It can be assumed as the real copper alloy that has produced in antiquity by adding Sn to Cu by different procedures to increase mechanical and chemical characteristics of copper. Tin make copper more fluidity and easier to cast like As and Zn, but about 10% tin, the metal would be harder and stronger than As and Zn addition (Maddin et al., 1977). Intentional alloying of copper with tin is usually considered to be indicative by a tin content of more than 1% or in some cases more than 5 percent.

Tin bronze (Cu-Sn) alloy became known in the late 4th millennium BCE and the beginning of the 3rd millennium BCE in Mesopotamia and western Iran (Khuzestan and Luristan regions). Nevertheless, the extensive consumption of tin and tin bronze in Mesopotamia emerges at middle of 3rd millennium BCE and in Iran even later (Thornton, 2009a).

The first appearance of producing the intentional bronze (Cu-Sn) alloy in Iran may be occurred in Kalleh Nissar graveyard in Luristan, western Iran at late 4th millennium BCE (Thornton, 2009a; Fleming et al., 2005). There are the evidences that direct mixing difference ores has used to create a single alloy in Kalleh Nissar artefacts by combining local copper and tin ores from the recently reported Deh Hosein deposit (Nezafati et al., 2006; Thornton, 2009a). The appearance of copper-tin alloys in Luristan about 3200–2800 BCE is a surprising phenomenon because the complete lack of even minor amounts of tin at many of other Iranian sites has observed until the end of the 3rd millennium BCE (Thornton, 2009a).

Of course, tin bronze examples have identified in other important sites such as Susa, Sialk, Tappeh Giyan, and Tappeh Yahya in Iran and Mundigak in Afghanistan at the early and mid-3rd millennium BCE (Thornton et al., 2002; Lamberg-Karlovsky, 1967). Even tough, an archaeometallurgical study on the site of Malyan, Fars region, showed the use of tin bronze in Middle Bronze Age of Iran between 2200-1800 BCE (Pigott et al., 2003).

In many texts, tin amount lower than 2-3 percent may be due to entering impurities from ores, but in higher amounts intentional alloying has occurred to produce tin bronze. Coghlan (1975) described four probably ways to making bronze:

1. Melting metallic Cu and metallic Sn with together as a mixture.
2. Adding (reducing) cassiterite (SnO_2) to melted copper under charcoal cover in crucible.
3. Smelting a natural copper-tin ore.
4. Smelting a mixture of a copper ore with together cassiterite (SnO_2).

In first procedure, by adding metallic tin to metallic copper, tin acts as deoxidant and increase fluidity of melt and casting ability. Also, adding metallic tin cause to decrease melting point of copper and it lowers by adding more tin (Pigott et al., 2003). The reducing SnO_2 in molten copper (second procedure) occurs at about 1200 °C in presence of charcoal in the crucible charge. The charcoal covered metal helps to maintain reducing conditions in crucible (Pigott et al., 2003; Coghlan, 1975). The third and fourth are named mixed smelting (or co-smelting) and has reported in relation with Luristan Bronzes (Nezafati et al., 2006).

Determination of tin resources in ancient time is one of the main problems and questions concerning of the starting bronze metallurgy in Bronze Age (3300-1500 BCE) in southwest Asia (Pigott, 2004a). It has long been discussed about tin sources for this huge amount of bronze production in Iranian Plateau in a long period of time from mid-Bronze Age to the end of Iron Age (Muhly, 1985; Maddin et al., 1977; Pigott et al., 2003; Coghlan, 1975).

It should be considered that the earliest tin sources might have probably been also copper ones. It is possible that the first bronze makers gained it accidentally. Probably at first a tin bearing copper ore had been smelted for production of metallic copper (or arsenical copper), but then because of the presence of tin in ore, bronze has been made accidentally. The final product of such process (arsenical copper and bronze) was identified by its golden color, its hardness and better casting properties. On the other hand it is possible that early smelters did not know cassiterite as an economical ore. They just knew that the ore of some specific mines result in better quality for metal products (Muhly, 1985).

One of the strongly probable sources of tin for Iran in ancient time was Afghanistan. There are many resources and mines of tin in this region and the probably cultural relationship between Iran and Afghanistan regions in prehistoric time has caused to tin sources in Afghanistan and Central Asia would be under considerations as main tin source for bronze metallurgy in Iran more than tin sources in Anatolia (Pigott et al., 2003; Muhly, 1985; Fleming et al., 2005). Of course late P. R. S. Moorey (1982) suggests two potential tin sources in Iranian Plateau: the Central Lut, a desert in central Iran, for which there is some minimal evidence, and a west or northwest Iranian source, maybe in Azerbaijan, for which there is still no special evidence. As noted above, Nezafati et al (2006) consider the mining region of Deh Hosein in Zagros Mountain, west central Iran as a source of tin for making bronze artefacts from Bronze to Iron Age, by combining local copper and tin ores and smelting them to produce bronze alloy (co-smelting).

Apart from sources of tin to use in bronze production, application of Cu-Sn alloy continued from Bronze Age to Iron Age in Iran. There are many examples from bronze production and use for making tools and artefacts in Iranian Iron Age, besides iron producing. Among copper-base artefacts from Iron Age sites that have been analyzed, bronze is the most common alloy, especially in western and northwestern Iran (Moorey, 1982). As a result, it is assumed that most Iron Age copper-base artefacts were of bronze and that the use of arsenical copper had waned considerably. Although, arsenical copper continued to be produced in this period and was excavated at Dailaman and Gilan regions (Pigott, 2004b).

The archaeological and experimental studies in different sites of Iron Age proved extensive archaeometallurgical finds, especially bronze artefacts. Examination of the archeological evidence suggests that the Gilan region by the Caspian Sea's littoral has been a bronze making center (Haerinck, 1988). In Marlik, the ancient site in northern Iran that is dated to late 2nd/early 1st millennium BCE, many bronze artefacts (figure 4) have found in graves beside of gold and silver vessels and decorative potteries. These are consisting of various classes of decorative artefacts such as spouted vessels, small statues of animals like deers, swords and daggers and etc (Negahban, 1999, 1996).

Fig. 4. Two bronze artefacts from Marlik graves, Left: a spouted vessel decorated with lion reliefs, Right: fantastic statue of a deer (Negahban, 1999).

At Hasanlu, an important Iron Age site in northwest Iran, more than 2,000 copper and bronze artefacts in the major various categories has found in archaeological excavations. Many of these artefacts have ornamented and decorated (Pigott, 1990). On the other hand, many bimetallic (two part artefacts made by bronze and Iron) artefacts have found such as pins with bronze lion head (Figure 5), spears with iron sockets and bronze blades, iron daggers with bronze cast-on hilts, and repoussé tin-bronze belt plaques with iron rivets (Pigott, 2004b, Thornton & Pigott, 2011).

Fig. 5. Three bronze artefacts from Hasanlu, Northwest Iran, 1st millennium BCE, a) Lion statue, b) decorative ring, c) bimetallic pin with bronze lion statue and iron pin shaft, Photos: M. Charehsaz and M. Ahmadi.

The Luristan Bronzes are one of the significant bronze collections from Iron Age of Iran. The name of "Luristan Bronzes" introduces a series of decorated bronze artefacts in a specific local style dating from the Iron Age (about 1500/1300 to 650/550 BCE) that belong to the geographical region of Luristan, central western Iran. These artefacts became known through large-scale illicit excavations started in the late 1920s, but their cultural context and provenance remained unspecified for a long time (Overlaet, 2006, 2004; Muscarella, 1990). Of course some controlled excavations has done in Luristan Region by various archaeologists such as late E. F. Schmidt in Surkh Dum site (Schmidt et al., 1989), Belgian Archaeological Mission in Iran (BAMI) under supervision of the late L.Vanden Berghe in Pusht-i Kuh Iron Age sites, western Part of Luristan (Overlaet, 2005; Muscarella, 1988) and M. Malekzadeh and A. Hassanpour in Sangtarashan (Oudbashi et al., in press) in Pish-i Kuh Region, Eastern Part of Luristan.

The Luristan bronzes are various in shapes and include lost wax casts as well as sheet metal objects consisting of different categories such as Horse gear includes horse-harness trappings and horse bits with decorative cheek pieces, arms and equipment include spiked axe heads, adzes, daggers, swords, whetstone handles, and quiver plaques, jewelries include rings, bracelets, pendants, and pins with cast or hammered sheet metal heads, an important series so-called "idols," also labeled "finials" or "standards," placed on tubular stands (Figure 6) and sheet metal vessels and jars (Overlaet, 2006, 2004; Muscarella, 1990, 1988; Moorey, 1974, 1971).

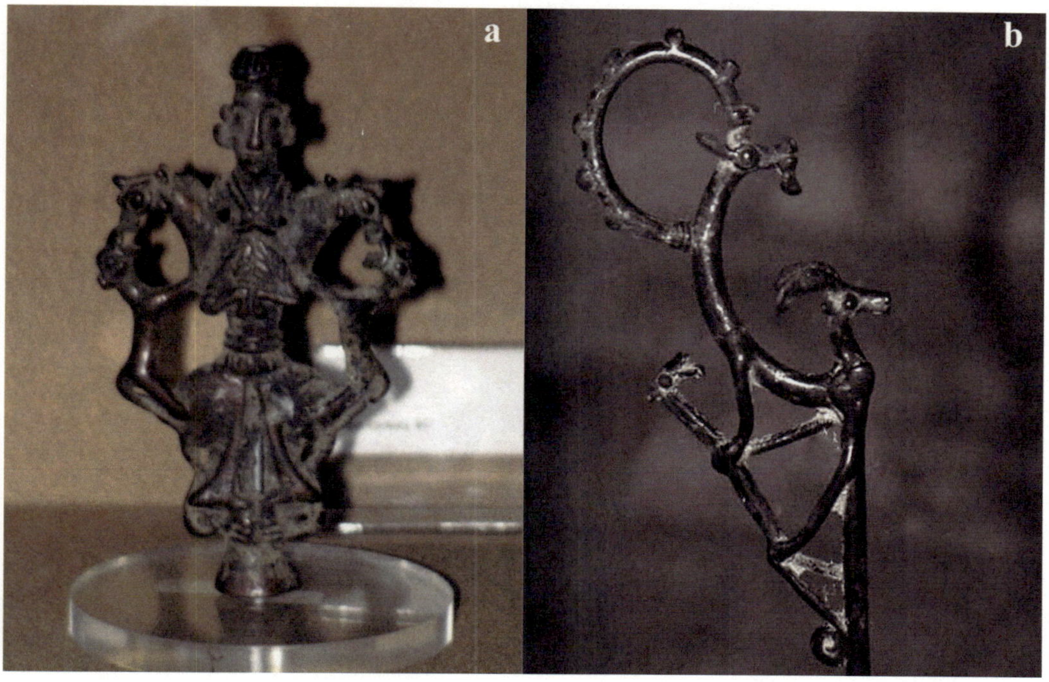

Fig. 6. a) A "Master of Animals" Standard from Luristan, with detached man head held by 4 animals. b) The upper part of an Animal Headed Pin, Luristan that shows combination of two goats and a felidae. Falakolaflak Museum, Khorramabad, Iran, Photo: O. Oudbashi.

The metallurgy in Elamite period was an important industry for producing various tools and artefacts especially religious and non-religious sculptures. There are many evidences from different parts through Elamite period by which the usage of copper alloys and especially bronze as well as craftsmanship of Elamite metalworkers will be obvious. One of the most significant examples of metallic sculpture from Middle Elamite period is the life-size statue of queen Napir-Asu discovered from Acropole of Susa that has dated to 14th century BCE (Figure 7). It is a bronze sculpture that has been casted in one piece and is 129 cm in height (without head). This impressive sculpture shows the ability and craftsmanship of Elamite master metalworkers in work with bronze and casting techniques. Another example is a the three dimensional representation bronze model called the Sit-Shamshi (Sunrise), from Acropole of Susa, 12th century BCE, 60 cm in Length (Figure 8) (Harper et al., 1992, Potts, 1999).

Fig. 7. Elamite Life-size bronze statue of queen Napir-Asu discovered, Susa, 14th century BCE, (Harper et al., 1992).

Fig. 8. Bronze model, Sit-Shamshi (Sunrise), Susa, 12th century BCE, (Harper et al., 1992).

4. Experimental study on Iranian bronzes

The study of bronze technology and the examination of copper based alloy artefacts from prehistoric time is one of the more interesting subjects in archaeometallurgical investigations around the world from some decades ago (Thornton et al., 2005; Paulin et al., 2003; Giumlia-Mair, et al., 2002; Thornton et al., 2002; Laughlin & Todd, 2000; Çukur & Kunça, 1989; Moorey, 1964). For technical studies, some metallic artefacts belonging to two Iranian archaeological sites, Haft Tappeh and Sangtarashan, have investigated and the results will be presented here. The aim of this investigation is to show metal production in these sites and a comparative study between metallurgical aspects of making copper based alloys. The investigations have done in two individual research projects and their results will be presented and compared here showing two specific metallurgical processes to make bronze and copper artefacts.

4.1 Description of sites

The investigated sites are among important archaeological finds in Iranian Plateau. The Haft Tappeh ancient site is located in Southwest Iran, in the central part of Khuzestan Province. Based on excavations done by late E. O. Negahban between 1965-1978, this site is belonging to early phase of Middle Elamite period (14th century BCE) and is remains an Elamite city named Kapanak, capital of Elamite's King Tepti Ahar (Negahban, 1993; Mofidi Nasrabadi, 2004; Potts, 1999). In the course of these excavations, many archaeological finds such as architectural remains, pottery, stone, bone, terracotta and metal artefacts are discovered (Negahban, 1993; Potts, 1999). Among these finds there are many metallurgical materials and remnants such as a two part furnace for making pottery and smelting ores, slag, matte as well as metal artefacts such as arrowheads, knives, swords, nails and pins, axes and other objects. Many of these findings were found in a workshop located at the east of excavated area (Negahban, 1993).

The recent archaeological activities in the field of Luristan Bronzes are excavations in Sangtarashan site between 2006-2011 by M. Malekzadeh and A. Hassanpour. This site is located in western Iran, Luristan region (Pish-i Kuh). The site is consisting of one cultural and historical layer belonging to Iron Age and type of discovered bronze collection suggests that this site can be dated to Iron Age IIA-B of western Iran about 1000-800 BCE (Oudbashi et al., in press; Overlaet, 2005; Overlaet, 2004; Azarnoush & Helwing, 2005). The Sangtarashan Bronze collection includes swords and daggers, axes, arrowheads, round vessels, spouted vessels, cups, decorative and ceremonial finials, decorative plaques, some bi-metallic (bronze-iron) artefacts and so on. This collection consist of individual group of excavated Luristan Bronzes, in high amount of founds as well as variety of manufacturing bronze artefacts. They have manufactured in a high level of craftsmanship and artistic skill (Oudbashi et al., in press).

4.2 Experimental

The experiments have designated to identification of chemical composition and microstructure of metallic artefacts from two sites. For this reason, 10 samples from different excavated metal pieces of Haft Tappeh and 10 samples from broken vessels of Sangtarashan (Figure 9) have selected and analyzed. To indentify chemical composition of samples, they analyzed by semi-quantitative chemical analysis with Scanning Electron Microscopy with

Energy Dispersive X-ray Spectrometry (SEM-EDS)[1] (Pollard et al., 2006; Pollard & Heron, 1996; Olsen 1988).

Metallographic samples were prepared by cutting a small piece from each sample, and then they mounted, polished and investigated through the optical microscope as well as SEM. It helps us to characterize microstructure of metal with its various details and aspects via high magnifications (Scott, 1991; Smith, 1975; Norton, 1967). To reveal microstructure and also for explaining the manufacturing methods, mounted samples have etched in ferric chloride ($FeCl_3$+HCl solution in H_2O) and examined under metallographic microscope (Caron et al., 2004; Scott, 1991).

4.3 Results and discussion

4.3.1 Alloy composition

The SEM-EDX results of 10 samples from different metallic artefacts and pieces belonging to Haft Tappeh are presented in Table 2. The major element in all samples is Cu and varies from about 67 to 95 percent in weight.

Fig. 9. left) 10 copper alloy samples investigated from Haft Tappeh, Middle Elamite Period, Right) One of the broken vessels from Sangtarashan Iron Age site, No. ST.08.

Other metallic elements in minor amount are Pb, Sn, Ni and Zn where have detected in an amount higher than 1% in weight. The elements such as Ag and Fe are lower than 1 percent and variable in content and may be considered as impurities from smelted ores. On the other hand, Al, Cl, Mg, S and Si show a minor amount in composition that they may be originated from soil, with the exception of S that can have another source, main ore.

[1] The SEM-EDS analyses were performed by 1) XL30 model, Philips in the SEM-EDS laboratory of Tarbiat-e Modarres University, Tehran, Iran, and 2) TESCAN model VEGA II, with a RONTEC BSE detector in SEM-EDS laboratory of RMRC, Tehran, Iran.

	Al	Ag	Cl	Cu	Fe	Mg	Mn	Ni	Pb	S	Sb	Si	Sn	Zn
HT.31	0.79	0.39	0.16	92.82	0.29	0.97	0.20	1.04	1.15	0.34	–	0.93	0.46	0.12
HT.35	0.94	0.41	0.39	93.20	0.32	0.98	0.19	0.94	–	0.39	0.11	0.59	0.77	0.77
HT.37	0.57	0.64	0.36	91.37	0.44	0.80	0.27	0.79	2.25	0.44	0.33	0.54	0.51	0.70
HT.43	–	–	–	93.26	0.12	–	–	0.77	4.97	–	–	0.20	0.25	0.43
HT.44	0.47	0.45	0.12	87.20	0.25	0.80	0.08	0.38	2.14	0.30	0.79	0.67	5.50	0.85
HT.47	0.17	–	6.87	67.27	0.39	0.45	0.16	0.56	3.98	0.60	–	1.06	17.96	0.10
HT.53	0.96	0.48	14.34	75.44	0.38	0.35	0.09	0.57	0.89	1.01	–	0.66	3.20	0.72
HT.54	1.08	0.26	0.17	68.39	0.15	2.03	0.02	4.76	18.22	2.02	0.33	1.77	0.15	0.47
HT.56	–	0.15	0.09	94.30	0.27	0.32	0.11	0.87	1.20	–	0.22	1.06	0.29	1.10
HT.71	0.32	0.20	0.11	95.16	0.44	0.40	0.12	0.39	0.61	0.38	0.33	1.04	0.15	0.35

Table 2. Chemical composition of Haft Tappeh Metallic Samples by SEM-EDS method in wt.%.

The samples are consisting of copper (HT.35, 71), leaded copper (HT.31, 37, 43, 54, 56) and leaded tin bronze (HT.44, 47, 53). As a matter of fact, nickel shows a significant role in samples No. HT.31 and HT.54 and zinc also in No. HT.56. Based on the results it can be suggested that the copper alloying in the analyzed Haft Tappeh artefacts has not been done by similar and controlled procedures and different compositions can be identified at them, although leaded copper and bronze are apparent in samples. The Pb is detected in 9 samples and it may be recommended that the alloyed samples are leaded with exception of 3 copper artefacts: HT.35, HT.53, and HT. 71 (lower than 1 percent).

The variety of alloy composition and determining different alloys, various amount of alloying elements in similar classes (such as lead and tin in bronzes) and presence of metallic elements such as Ni in minor content shows that the metalworking in Haft Tappeh may not followed from a specific process and an uncontrolled smelting and alloying procedure has been applied in this period of time.

On the other hand, the ore sources of Elamite metallurgy has not been identified yet and it can be state that the different metal sources may be used to metal smelting in this site.

Table 3 shows the results of SEM-EDS analysis performed on 10 broken vessels from Sangtarashan Iron Age site. The results determine that only two elements play main role to form alloy composition in sample: Cu and Sn. The tin content is variable from 7.75 to 13.56% and cannot consider as an alloying pattern based on Sn.

Other elements such as Ag, As, Fe, Ni, P, Pb, Sb and Zn are lower than 1% in many samples and don't play an important role in alloying process. They can be considered as impurities that may originated from mother stone ores.

	Ag	As	Cu	Fe	Ni	P	Pb	Sb	Sn	Zn
ST.01	–	0.02	88.61	–	–	0.03	–	–	11.32	–
ST.02	0.67	0.03	90.79	–	–	0.14	0.61	–	7.75	0.01
ST.03	0.54	0.73	86.18	0.42	0.42	0.24	0.81	–	9.43	1.23
ST.04	0.82	1.18	82.86	0.10	–	0.15	0.56	–	13.56	0.77
ST.05	–	0.03	90.40	0.01	–	0.03	–	–	9.51	0.01
ST.07	0.48	0.04	87.40	0.32	0.41	0.13	1.23	–	9.60	0.41
ST.08	0.56	0.03	86.83	0.23	–	0.17	0.26	–	11.63	0.29
ST.09	0.81	0.04	83.82	0.40	–	0.29	0.43	–	12.78	1.43
ST.10	0.27	0.03	90.59	–	–	0.10	0.24	–	8.76	0.01
ST.11	–	0.22	89.28	–	–	0.07	0.24	0.37	9.79	0.02

Table 3. Chemical composition of Sangtarashan Metallic Samples by SEM-EDS method in wt.%.

4.3.2 Microstructure

The microstructures of the unetched samples of both sites observed by optical microscope (OM) and SEM consist of many fine gray-green inclusions spread in the metallic matrix (Figure 11a, 12a). To identify chemical composition of these inclusions, three samples from each site has analyzed by SEM-EDS microanalysis. The results of main elements detected in analyses are presented in Table 4.

In Table 4, only elements detected more than 1% wt. in amount are presented. The major elements detected in all samples are Cu and S. Also, Fe, Sn, Si and Pb are available in minor content in samples. The elemental composition of inclusions suggests that the inclusions are composed of copper sulfides with some iron sulfides. Sn (and Pb in sample HT.37) constituents probably associated to the bronze matrix and have no relation with the composition of inclusions.

	Cu	Fe	Pb	S	Si	Sn
HT.35	77.88	1.25	–	14.79	–	–
HT.37	75.50	2.08	1.30	15.04	1.30	–
HT.44	79.03	-	–	15.19	–	1.13
ST.03	83.48	1.10	–	12.48	–	2.94
ST.05	79.50	0.95	–	18.44	–	1.11
ST.08	79.21	2.77	–	15.77	–	2.25

Table 4. Results of SEM-EDS analysis of inclusions for three samples from each sites (wt.%). The elements with higher amount than 1% are presented.

Figure 10 presents ternary phase diagram of the Cu-Fe-S system based on SEM-EDS analysis of inclusions. The proportion of Cu/S and location of points in Figure 10, strongly suggests the presence of chalcocite (Cu_2S) or digenite (Cu_9S_5). Also, with regard to the appearance of low to variable amounts of Fe in the composition, this may reflect the existence of iron sulfides like Pyrrhotite ($Fe_{1-x}S$) (Klein & Hurlbut Jr., 1999). In fact, the composition of inclusions suggests they are residues of sulfidic ores (Singh & Chattopadhyay, 2003; Bachmann, 1982). Sulfidic inclusions have been observed in many copper alloy artefacts from other examined sites in prehistoric Iran such as Luristan Iron Age sites (Fleming et al., 2006; Fleming et al., 2005) and some Chalcolithic and Bronze Age sites in southeastern Iran (Thornton & Lamberg-Karlovsky, 2004).

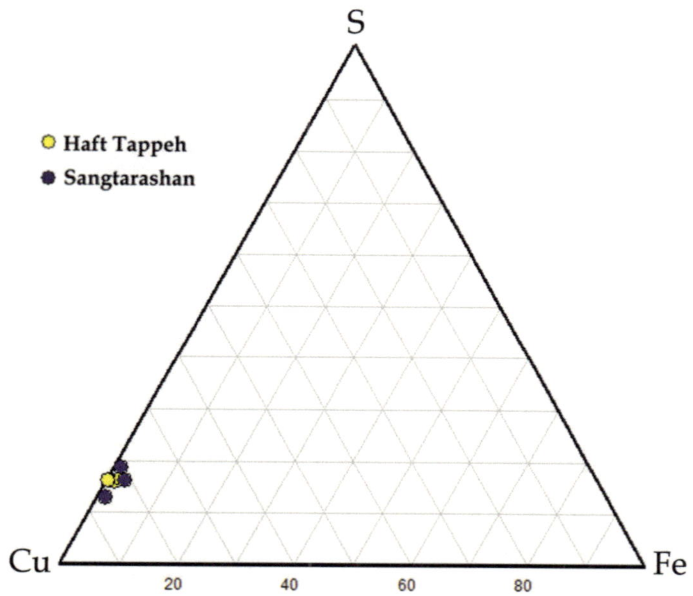

Fig. 10. Ternary phase diagram of Cu-Fe-S system with respect to the inclusions' composition. The main component is copper sulfide.

To identify microstructure in detail and to make clear the manufacturing method of the artefacts, three mounted samples from each site were etched in ferric chloride solution and examined under metallographic (optical) microscope (Caron et al., 2004; Scott, 1991). The etched microstructure of samples HT.31, HT. 43 and HT. 44 have studied from Haft Tappeh. The microstructure is consisting of worked and recrystallized grains of α solid solution of copper with twinning lines in grains (Figure 11b-d).

FCC metals such as copper (α solid solution in Cu-Sn system) recrystallize by a twinning process (except for aluminum). New crystals that grow following the annealing after cold-working in copper and its alloys produce the effect of a mirror reflection plane within the crystals, with the result that parallel straight lines can be seen in etched sections traversing part or all of the individual grains of the metal. The straight twin lines shows that the final stage of shaping operation has been annealing but the deformed (curved) twin lines suggest that the metal has been worked finally and if heavily worked, the slip lines can be seen in grains (Scott, 1991).

Fig. 11. Microstructure of Haft Tappeh samples: a) SEM micrograph of an inclusion, Sample HT.37, b) microstructure of sample HT.43, c) sample HT.44, d) sample HT. 31. All are etched in ferric chloride solution and show equi-axed grains with twinning.

The etched microstructure of Sangtarashan samples (ST.01, ST.03, ST.04) is even more similar than Haft Tappeh ones and of course shows worked and recrystallized grains of α solid solution (Figure 12b-d). The twinning microstructure in these samples also occurred due to hammering and annealing of cast bronze and demonstrates that ancient metalworkers shaped the vessels by hammering as well as subsequent heat treatment. This process of hammering could have been used to shape a cast bronze ingot into sheet metal vessel, and in the course of the mechanical working the sheet metal was work-hardened. To remove this work hardening, (like Haft Tappeh samples) the metal has treated with heat between 500 to 800 °C (i.e. for copper alloys). This annealing process returns workability to the sheet metal (Siano et al 2006; Caron et al 2004; Scott 1991).

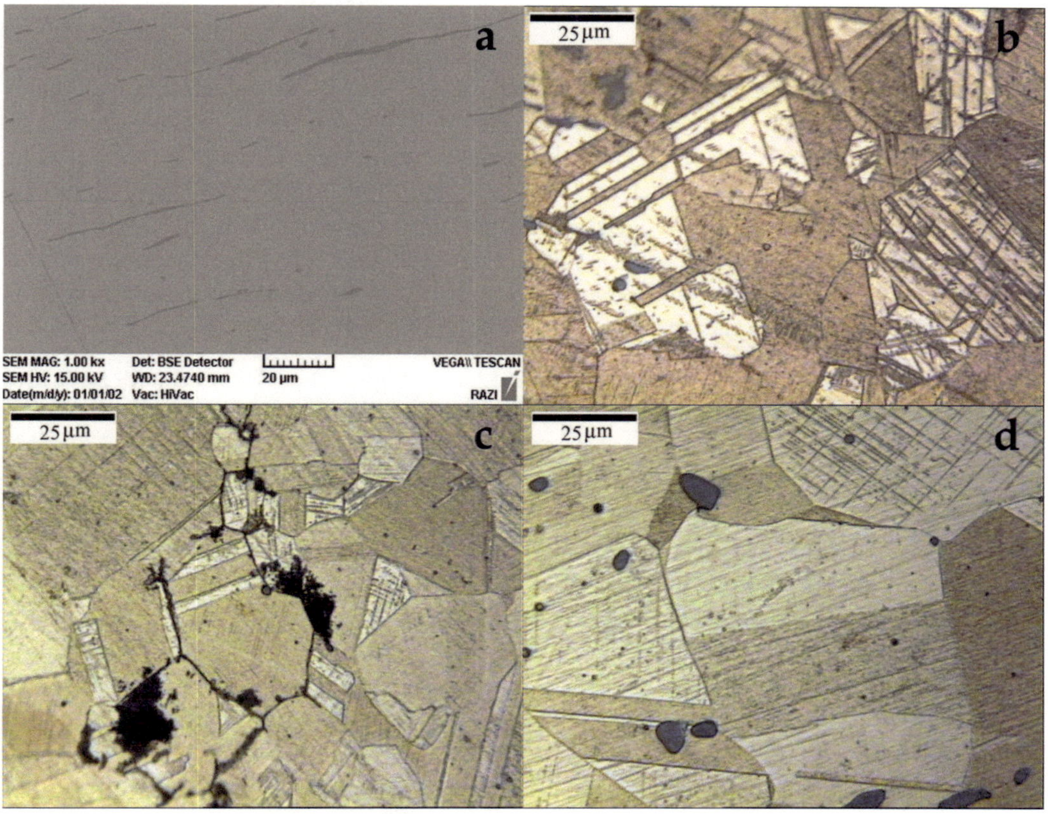

Fig. 12. Microstructure of Sangtarashan samples: a) SEM micrograph of elongated inclusions, Sample ST.08, b) microstructure of sample ST.01, c) sample ST.03, d) sample ST.04. All are etched in ferric chloride solution and show equi-axed grains with twin and slip lines.

5. Conclusion

Iran is a considerable region in archaeological researches and events and about 150 years has been taken into attention of many archaeologists, historians and scientist from worldwide. In fact, Iranian Plateau was one of the pioneer regions in many aspects of technological and manufacturing history from ancient times. The investigations and researches on the history of metallurgy and metalworking states that the earliest experiences and events in metal usage in the world have occurred in Anatolia, Caucasus, Iran and Levant.

Copper is the first metal for making tools and ornaments in Iran (like other regions). The Archaeometallurgy of copper can be classified in four main stages in Iran prehistory. Initially, the ancient metalworkers have used native copper to making artefacts from about 9000 years ago. The second stage has been melting native copper and casting it to specific shapes in early Chalcolithic period. Some of native copper resources in Iran contained arsenic and it caused to occurrence of early accidental alloying and emergence of arsenical copper. The third stage is smelting of copper from ores that is started by smelting oxidic copper ores such as cuprite and malachite and then polymetallic sulfidic ores in Chalcolithic; and consequently, many of objects made by smelted copper also have arsenic as an impurity or alloying elements. In last stage, by progressing metallurgical experiences in prehistory, production of a new alloy has occurred in the late 4th millennium BCE (early Bronze Age). Tin bronze has observed about 3000 BCE but has been common about 1000 years later and even in Iron Age (1500-550). Bronze usage in Iron Age is an important stage of copper alloys metallurgy in Iran. There are many examples of decorative bronze collections and artefacts from Iron Age sites especially in graves. These are discovered in many Iron Age graves and may assume as vows religious cemeterial objects, such as Luristan and Marlik bronze collections. Bronze production has been continued in the next periods although some changes in metallurgical traditions have occurred in that time.

The examination of some metallic samples from Haft Tappeh (Middle Elamite period) and Sangtarashan (Iron Age) sites shows differences and similarities in alloying and shaping processes. The main difference between these sites is alloying. Haft Tappeh shows a variable alloying process to make copper alloys. There are three significant composition in samples: copper, leaded copper and leaded tin bronze. The considerable subject in all samples is presence of tin, even in low amount. This may cause to assume all samples as bronze (low tin bronze and tin bronze). Nevertheless, presence of Sn may be due to application of copper tin-bearing ores. On the other hand, alloy composition in Sangtarashan shows an apparent pattern in bronze production; although tin content in samples is variable and it state that the alloying has not done in a controlled condition. Presence of copper sulfide inclusions in microstructure suggests the employed ores in both sites were copper sulfidic ores. Low amount of As in both and Pb in Sangtarashan are two interesting aspects of alloy composition and may related to ore sources, although need to be considered as a research subject in the future. The microstructure is consisting of equi-axed and worked grains of copper solid solution explaining the shaping method has been cold working on casted or smelted copper or bronze ingot, continuing with annealing as a stage to remove work hardening, especially in thin vessels of Sangtarashan.

Despite of history of metallurgy in Iran, especially bronze, many investigations about Iran archaeometallurgy only focus on archaeological subjects and relations and connections between cultures and civilizations and the technical studies are very limited in compare of archaeological ones. It seems to need to a large scale study on metallurgical technology in ancient and historic Iran to reveal various aspects of archaeometallurgical activities in Iran, especially with regard to bronze technology.

6. Acknowledgements

We would like to thank to A. Hassanpour and M. Malekzadeh from archaeological Expedition of Sangtarashan and H. Fadaei from Haft Tappeh Conservation Project for their helps to access to metallic samples, B. Rahmani, K. Asgari, M. Ghadrdan and S. Ali Asghari from RMRC, and A. Rezaei from Tarbiat-e Modarres University of Tehran and M. Ghobadi from Art University of Isfahan for their helps to carry out SEM-EDS analyses and OM observations, Dr. C. P. Thornton, University of Pennsylvania Museum, USA, for his valuable helps, comments and information, Atefeh Shegofteh, Art Conservator, for her helps to preparing samples, text and illustrations, M. Charehsaz, Tabriz Islamic Art University and M. Ahmadi ICHTO of Western Azerbaijan for their helps about Hasanlu photos, and Dr. H Ahmadi, Faculty of Conservation, Art University of Isfahan and Dr. A. Nazeri, Research Deputy of Art University of Isfahan for their helps to make opportunities for publishing this chapter.

7. References

Arab, R. & Rehren, T. (2004). The Pyrotechnological Expedition of 1968, In *Persia's Ancient Splendour, Mining, Handicraft and Archaeology,* Stöllner T., Slotta R. & Vatandoust A. (eds.), pp. (550-555), Deutsches Bergbau-Museum, Bochum.

Azarnoush, M. & Helwing, B. (2005). Recent Archaeological Research in Iran - Prehistory to Iron Age. *Archaeologische Mitteilungenaus Iran und Turan,* Vol. 37, pp. (189-246).

Bachmann, H. G. (1982). *The Identification of Slags from Archaeological Sites,* Occasional Publications, No. 6, Institute of Archaeology, UCL, London.

Benoit, A. (2004). Susa, In *Persia's Ancient Splendour, Mining, Handicraft and Archaeology,* Stöllner T., Slotta R. & Vatandoust A. (eds.), pp. (178-192), Deutsches Bergbau-Museum, Bochum.

Bernbeck, R. (2004). Iran in the Neolithic, In *Persia's Ancient Splendour, Mining, Handicraft and Archaeology,* Stöllner T., Slotta R. & Vatandoust A. (eds.), pp. (140-147), Deutsches Bergbau-Museum, Bochum.

Birmingham, J., Kennon, N. F. & Malin, A. S. (1964). A "Luristan" Dagger: An Examination of Ancient Metallurgical Techniques. *Iraq,* Vol. 26, No. 1, pp. (44-49).

Caron, R. N., Barth, R. G. & Tyler, D. E. (2004). Metallography and Microstructures of Copper and its Alloys, In *ASM handbook,* Metallography and Microstructures, Vol. 9, pp. (775-788), ASM International, Materials Park, Ohio.

Chegini, N. N., Helwing, B., Parzinger, H. & Vatandoust, A. (2004). A Prehistoric Industrial Settlement on the Iranian Plateau–Research at Arisman, In *Persia's Ancient Splendour, Mining, Handicraft and Archaeology,* Stöllner T., Slotta R. & Vatandoust A. (eds.), pp. (210-216), Deutsches Bergbau-Museum, Bochum.

Coghlan, H. H. (1975). Notes on the Prehistoric Metallurgy of Copper and Bronze in the Old World, Occasional Paper on Technology 4, 2nd Ed., Oxford.

Çukur, A. & Kunça, S. (1989). Development of Bronze Production Technologies in Anatolia. *Journal of Archaeological Science,* Vol. 16, pp. (225-231).

Dougherty, R. C. & Caldwell, J. R. (1966). Evidence of Early Pyrometallurgy in the Kerman Range in Iran, *Science,* Vol. 153, pp. (984-985).

Dyson Jr., R. H. & Voigt, M. M. (1989), Bronze Age, In *Encyclopedia Iranica,* Yarshater E. (ed.), Vol. IV, pp. (472-478), Routledge & kegan Paul, London and New York.

Emami, S. M. & Oudbashi, O. (2008). Comparative Analysis on Polymetallic Ore Sources with Respect on Archaeometallurgical Melting Process in Lasoleyman Area closed by Meymand Cave Dwelling Village in Kerman, Central Iran, *Proceeding of International Conference of Ancient Mining in Turkey and Eastern Mediterranean (AMiTEM 2008),* Yalçin, Ü., Özbal, H. & Pasamemetoglu, A. G. (Eds.), pp. (351-369), June 15-22, 2008, ATILIM university Publications, Ankara.

Fleming, S. J., Pigott, V. C., Swann, C. P., Nash, S. K., Haerinck, E. & Overlaet, B. (2006). The Archaeometallurgy of War Kabud, Western Iran. *Iranica Antiqua,* Vol. XLI, pp. (31-57).

Fleming, S. J., Pigott, V. C., Swann, C. P. & Nash, S. K. (2005). Bronze in Luristan: Preliminary Analytical Evidence from Copper/bronze Artifacts Excavated by the Belgian Mission in Iran. *Iranica Antiqua,* Vol. XL, pp. (35-64).

Frame, L. D. (2007). Metal finds from Godin Tepe, Iran: Production, Consumption, and Trade, MS thesis, Department of Materials Science and Engineering, University of Arizona.

Frame, L. D. (2004). Investigations at Tal-i Iblis: Evidence for Copper Smelting During the Chalcolithic Period, BS thesis, Department of Materials Science and Engineering, Massachusetts Institute of Technology.

Giumlia-Mair, A., Keall, E. J., Shugar, A. N. & Stock, S. (2002). Investigation of a Copper-based Hoard from the Megalithic Site of al-Midamman, Yemen: an Interdisciplinary Approach. *Journal of Archaeological Science,* Vol. 29, pp. (195–209).

Haerinck, E. (1988), The Iron Age in Guilan: Proposal for a Chronology, In *Bronzeworking Centres of Western Asia 1000-539 B.C.,* J. Curtis, (ed.), pp. (63-78), London.

Harper, P. O., Aruz, J. & Tallon, F. (Eds.) (1992). *The Royal City of Susa: Ancient Near Eastern Treasures in the Louvre,* The Metropolitan Museum of Art, New York.

Henrickson, E. F. (1992)a. Chalcolithic Era, In *Encyclopedia Iranica,* Yarshater E. (ed.), Vol. V, pp. (347-353), Mazda Publishers, California.

Henrickson, E. F. (1992)b. CERAMICS iv. The Chalcolithic Period in the Zagros Highlands, In *Encyclopedia Iranica,* Yarshater E. (ed.), Vol. V, pp. (278-282), Mazda Publishers, California.

Henrickson, E. F. (1985). An Updated Chronology of the Early and Middle Chalcolithic of the Central Zagros Highlands, Western Iran. *Iran,* Vol. 23, pp. (63-108).

Heskel, D. L. & Lamberg-Karlovsky, C. C. (1986). Metallurgical Technology, In *Excavations at Tepe Yahya, Iran: The Early Periods,* Lamberg-Karlovsky, C. C. & Beale, T., (Eds.), pp. (207-214), Cambridge.

Heskel, D. L. & Lamberg-Karlovsky, C. C. (1980). An alternative sequence for the development of metallurgy: Tepe Yahya, Iran, In *The coming of the Age of Iron,* T. Wertime & J. Muhly (Eds.), pp. (229–266), New Haven: Yale University Press.

Hole, F. (2008). Paleolithic Age in Iran, In *Encyclopaedia Iranica Online,* Yarshater E. (ed.), Originally Published: July 28, 2008 Available at http://www.iranica.com/articles/paleolithic-age-in-iran

Hole, F. (2004). Neolithic Age in Iran, In *Encyclopaedia Iranica Online,* Yarshater E. (ed.), Originally Published: July 20, 2004 Available at http://www.iranica.com/articles/neolithic-age-in-iran

Hole, F. (2000). New radiocarbon dates for Ali Kosh, Iran, *Neolithics,* Vol. 1, p. (13).

Klein, C. & Hurlbut Jr., C. S. (1999). *Manual of Mineralogy*, Revised 21st Edition (After J.D. Dana), John Wiley and Sons INC., Toronto.

Lamberg-Karlovsky, C. C. (1967). Archeology and Metallurgical Technology in Prehistoric Afghanistan, India, and Pakistan. *American Anthropologist*, Vol. 69, No. 2, pp. (145-162).

Laughlin, G. J. & Todd, J. A. (2000). Evidence for Early Bronze Age Tin Ore Processing. *Materials Characterization*, Vol. 45, pp. (269-273).

Maddin, R., Wheeler, T. S. & Muhly, J. D. (1977). Tin in the Ancient Near East: Old Questions and New Finds. *Expedition*, Vol. 19, No. 2, pp. (35-47).

Mofidi Nasrabadi, B. (2004). Elam: Archaeology and History, In *Persia's Ancient Splendour, Mining, Handicraft and Archaeology*, Stöllner T., Slotta R. &Vatandoust A. (eds.), pp. (294-308), Deutsches Bergbau-Museum, Bochum.

Moorey, P. R. S. (1964). An Interim Report on Some Analyses of "Luristan Bronzes". *Archaeometry*, Vol. 7, pp. (72-79).

Moorey, P. R. S. (1969). Prehistoric Copper and Bronze Metallurgy in Western Iran (With Special Reference to Lūristān). *Iran*, Vol. 7, pp. (131-153).

Moorey, P. R. S. (1971). *Catalogue of the Ancient Persian Bronzes in the Ashmolean Museum*, Oxford University Press, Oxford.

Moorey, P. R. S. (1974). *Ancient Persian Bronzes in the Adam Collection*, London.

Moorey, P. R. S. (1982). Archaeology and Pre-Achaemenid Metalworking in Iran: A Fifteen Year Retrospective. *Iran*, Vol. 20, pp. (81-101).

Muhly, J. D. (1985). Sources of Tin and the Beginnings of Bronze Metallurgy. *American Journal of Archaeology*, Vol. 89, No. 2, pp. (275-291).

Muscarella, O. W. (2006). Iron Age, In *Encyclopaedia Iranica Online*, Yarshater E. (ed.), Originally Published: December 15, 2006 Available at http://www.iranica.com/articles/iron-age

Muscarella, O. W. (1990). Bronzes of Luristan, In *Encyclopedia Iranica*, Yarshater E. (ed.), Vol. IV, pp. (478-483), Routledge & Kegan Paul, London & New York.

Muscarella, O. W. (1988). *Bronze and Iron: Ancient Near Eastern Artifacts in The Metropolitan Museum of Art*, Metropolitan Museum of Art, New York.

Negahban, E. O. (1999). *Marlik Excavations*, Vol. 1, ICHO Press, Tehran (in Farsi).

Negahban, E. O. (1996). *Marlik, the Complete Excavation Report*, University Museum Monograph 87, University of Pennsylvania, Philadelphia.

Negahban, E. O. (1993). *Excavations in Haft Tappeh, Khuzestan*, ICHO Press, Tehran (in Farsi).

Nezafati, N., Pernicka, E. & Momenzadeh, M. (2006). Ancient tin: Old question and a new answer. *Antiquity*, Vol. 80, p. (308).

Norton, J. T. (1967). Metallography and the Study of Art Objects, In *Application of Science in the Examination of Works of Art*, Young, W. J. (ed.), pp. (13-19), September 7-16, 1965, Research Laboratory of Museum of Fine Arts, Boston.

Olsen, S. L. (1988), Applications of Scanning Electron Microscopy in Archaeology, In: *Advances in Electronics and Electron Physics*, Vol. 71, pp. (357-380), Academic Press.

Oudbashi, O., Emami, S. M., Malekzadeh, M., Hassanpour, A. & Davami, P. (in press). Archaeometallurgical Studies on the Bronze vessels from "Sangtarashan", Luristan, W-Iran. *Iranica Antiqua*, Vol. XLVIII, (Accepted).

Oudbashi, O., Emami, S. M. & Bakhshandehfard, H. (2009). Preliminary Archaeometallurgical Studies on Mineralogical Structureand Chemical

Composition of Ancient Metal Objects and Slag from Haft Tepe, Southwest Iran, Khuzestan (Middle Elamite Period), *Proceedings of 36th International Symposium on Archaeometry, ISA 2006*, 2-6 May 2006, Moreau, J. F., Auger, R., Chabot, J. & Herzog, A. (Eds.), Université Laval, Quebec City, Canada, pp. (407-412), CELAT Publications.

Overlaet, B. (2006). Luristan Bronzes: I. The Field Research, In *Encyclopaedia Iranica Online*, Yarshater E. (ed.), Originally Published: November 15, 2006 Available at http://www.iranica.com/articles/luristan-bronzes-i-the-field-research.

Overlaet, B. (2005). The Chronology of the Iron Age in the Pusht-i Kuh, Luristan. *Iranica Antiqua*, Vol. XL, pp. (1-33).

Overlaet, B. (2004). Luristan Metalwork in the Iron Age, In *Persia's Ancient Splendour, Mining, Handicraft and Archaeology*, Stöllner T., Slotta R. & Vatandoust A. (eds.), pp. (328-338), Deutsches Bergbau-Museum, Bochum.

Paulin, A., Spaic´, S., Zalar, A. & Trampuž-Orel, N. (2003). Metallographic Analysis of 3000-Year-Old Kanalski Vrh Hoard Pendant. *Materials Characterization*, Vol. 51, pp. (205–218).

Pernicka, E. (2004). Copper and Silver in Arisman and Tappeh Sialk and the Early Metallurgy in Iran, In *Persia's Ancient Splendour, Mining, Handicraft and Archaeology*, Stöllner T., Slotta R. & Vatandoust A. (eds.), pp. (232-239), Deutsches Bergbau-Museum, Bochum.

Pigott, V. C. (2004)a. On the Importance of Iran in the Study of Prehistoric Copper-Base Metallurgy, In *Persia's Ancient Splendour, Mining, Handicraft and Archaeology*, Stöllner T., Slotta R. & Vatandoust A. (eds.), pp. (28-43), Deutsches Bergbau-Museum, Bochum.

Pigott, V. C. (2004)b. Hasanlu and the Emergence of Iron in Early 1st Millennium BC, Western Iran, In *Persia's Ancient Splendour, Mining, Handicraft and Archaeology*, Stöllner T., Slotta R. & Vatandoust A. (eds.), pp. (350-357), Deutsches Bergbau-Museum, Bochum.

Pigott, V. C. (1990). Bronze; in Pre-Islamic Iran, In *Encyclopedia Iranica*, Yarshater E. (ed.), Vol. IV, pp. (457-471), Routledge & kegan Paul, London and New York.

Pigott, V. C., Rogers, H. C. & Nash, S. K. (2003), Archaeometallurgical Investigations at Tal-e Malyan: The Evidence for Tin-Bronze in the Kaftari Phase, In *Yeki Bud, Yeki Nabud: Essays on the archaeology of Iran in honor of William M. Sumner*, Miller, N. F. & Abdi, K. (Eds.), pp. (161-175), University of Pennsylvania Museum of Archaeology and Anthropology, Philadelphia

Pigott, V. C., Howard, S. M. & Epstein, S. M. (1982). Pyrotechnology and Culture Change at Bronze Age Tepe Hisar (Iran), In *Early Pyrotechnology, The Evolution of the First Fire-Using Industries*, Wertime, T. A. & Wertime, S. A. (Eds.), pp. (215-236), Washington D. C..

Pleiner, R. (2004). Memories of the Archaeometallurgic Expeditions to Iran and Afghanistan in the Years 1966 and 1968, In *Persia's Ancient Splendour, Mining, Handicraft and Archaeology*, Stöllner T., Slotta R. & Vatandoust A. (eds.), pp. (556-560), Deutsches Bergbau-Museum, Bochum.

Pollard, A. M., Batt, C., Stern, B. & Young, S. M. M. (2006). *Analytical Chemistry in Archaeology*, Cambridge University Press, New York.

Pollard, A. M. & Heron, C. (1996). *Archaeological Chemistry*, The Royal Society of Chemistry, Cambridge.

Potts, D. T. (1999).*The Archaeology of Elam: Formation and Transformation of an Ancient Iranian State*, Cambridge University press, Cambridge.

Schmidt, E. F., Van Loon, M. N. & Curvers, H. H. (1989). *The Holmes expeditions to Luristan*, 2 Vols., The Oriental Institute of The University of Chicago Publications 108, Chicago.

Scott, D. A. (2002). *Copper and Bronze in Art: Corrosion, Colorants and Conservation*, Getty Conservation Institute Publications, Los Angeles.

Scott, D. A. (1991). *Metallography and Microstructure of Ancient and Historic Metals*, Getty Conservation Institute, Los Angeles.

Siano, S., Bartoli, L., Santisteban, J. R., Kockelmann, W., Daymond, M. R., Miccio, M. & De Marinis, G. (2006). Non-Destructive Investigation of Bronze Artefacts from the Marches National Museum of Archaeology Using Neutron Diffraction. *Archaeometry*, Vol. 48, pp. (77–96).

Singh, A. K. & Chattopadhyay, P. K. (2003). Carinated and knobbed copper vessels from the Narhan Culture, India. *IAMS*, Vol. 23, pp. (27-31).

Smith, C. S. (1975), Metallography-How it Started and Where it's Going. *Metallography*, Vol. 8, pp. (91-103).

Smith, C. S. (1967), The Interpretation of Microstructures of Metallic Artifacts, In *Application of Science in the Examination of Works of Art*, Young, W.J. (ed.), pp. (20-52), September 7-16, 1965, Research Laboratory of Museum of Fine Arts, Boston.

Tala'i, H. (2008)a. *The Iron Age of Iran*, The Organization for Researching and Composing University Textbooks in the Humanities (SAMT), First Edition, Tehran, (in Farsi).

Tala'i, H. (2008)b. *The Bronze Age of Iran*, The Organization for Researching and Composing University Textbooks in the Humanities (SAMT), Second Edition, Tehran, (in Farsi).

Tala'i, H. (1996). *Iranian Art and Archaeology in the First Millennium B.C.*, The Organization for Researching and Composing University Textbooks in the Humanities (SAMT), Tehran, (in Farsi).

Thornton, C. P. (2010), The rise of Arsenical Copper in Southeastern Iran. *Iranica Antiqua*, Vol. XLV, pp. (31-50).

Thornton, C. P. (2009)a. The Emergence of Complex Metallurgy on the Iranian Plateau: Escaping the Levantine Paradigm. *Journal of World Prehistory*, Vol. 22, pp. (301-327).

Thornton, C. P. (2009)b. The Chalcolithic and Early Bronze Age metallurgy of Tepe Hissar, northeast Iran: A challenge to the Levantine paradigm, PhD dissertation, Department of Anthropology, University of Pennsylvania.

Thornton, C. P. & Pigott, V. C. (2011). Blade-Type Weaponry of Hasanlu IVB, In *Peoples and Crafts of Hasanlu IVB*, de Schauensee M. & Dyson Jr., R. H. (eds.), University of Pennsylvania Museum Publications, Philadelphia, pp. (135-182).

Thornton, C. P. & Roberts, B. W. (2009). Introduction: The Beginnings of Metallurgy in Global Perspective. *Journal of World Prehistory*, Vol. 22, pp. (181-184).

Thornton, C. P., Rehren, T. & Pigott, V. C. (2009). The Production of Speiss (Iron Arsenide) During the Early Bronze Age in Iran. *Journal of Archaeological Science*, Vol. 36, pp. (308 - 316).

Thornton, C. P. & Rehren, T. (2007). Report on the First Iranian Prehistoric Slag Workshop. *Iran*, Vol. XLV, pp. (315-318).

Thornton, C. P., Lamberg-Karlovsky, C. C., Liezers, M. & Young, S. M. M. (2005). Stech and Pigott Revisited: New Evidence Concerning the Origin of Tin Bronze in Light of Chemical and Metallographic Analyses of the Metal Artefacts from Tepe Yahya, Iran, In *Proceedings of the 33rd International Symposium on Archaeometry*, 22-26 April 2002, Amsterdam, Kars, H. & Burke, E. (eds.), Geoarchaeological and Bioarchaeological Studies 3, pp. (395-398), Vrije Universiteit, Amsterdam.

Thornton, C. P. & Lamberg-Karlovsky, C. C. (2004). Tappeh Yahya and the Prehistoric Metallurgy of South-eastern Iran, In *Persia's Ancient Splendour, Mining, Handicraft and Archaeology*, Stöllner T., Slotta R. & Vatandoust A. (eds.), pp. (264-273), Deutsches Bergbau-Museum, Bochum.

Thornton, C. P. & Ehler, C. B. (2003). Early Brass in the Ancient Near East. *IAMS*, Vol. 23, pp. (3-8).

Thornton, C. P., Lamberg-Karlovsky, C. C., Liezers, M. & Young, M. M. (2002). On Pins and Needles: Tracing the Evolution of Copper-base Alloying at Tepe Yahya, Iran, via ICP-MS Analysis of Common-Place Items. *Journal of Archaeological Science*, Vol. 29, pp. (1451–1460).

Wertime, T. A. (1973). The Beginning of Metallurgy: A New Look. *Science*, Vol. 182, No. 4115, pp. (875- 887).